Lecture Notes in Mechanical Engineering

Lecture Notes in Mechanical Engineering (LNME) publishes the latest developments in Mechanical Engineering—quickly, informally and with high quality. Original research or contributions reported in proceedings and post-proceedings represents the core of LNME. Volumes published in LNME embrace all aspects, subfields and new challenges of mechanical engineering.

To submit a proposal or request further information, please contact the Springer Editor of your location:

Europe, USA, Africa: Leontina Di Cecco at Leontina.dicecco@springer.com
China: Ella Zhang at ella.zhang@cn.springernature.com
India, Rest of Asia, Australia, New Zealand: Swati Meherishi at swati.meherishi@springer.com

Topics in the series include:

- Engineering Design
- Machinery and Machine Elements
- Mechanical Structures and Stress Analysis
- Automotive Engineering
- Engine Technology
- Aerospace Technology and Astronautics
- Nanotechnology and Microengineering
- Control, Robotics, Mechatronics
- MEMS
- Theoretical and Applied Mechanics
- Dynamical Systems, Control
- Fluid Mechanics
- Engineering Thermodynamics, Heat and Mass Transfer
- Manufacturing Engineering and Smart Manufacturing
- Precision Engineering, Instrumentation, Measurement
- Materials Engineering
- Tribology and Surface Technology

Indexed by SCOPUS, EI Compendex, and INSPEC.

All books published in the series are evaluated by Web of Science for the Conference Proceedings Citation Index (CPCI).

To submit a proposal for a monograph, please check our Springer Tracts in Mechanical Engineering at https://link.springer.com/bookseries/11693.

Magdalena Hryb · Jiehan Zhou ·
Michał Wieczorowski · Ewa Dostatni ·
Piotr D. Moncarz
Editors

Advances in Manufacturing V

Volume 4 - Measurement and Control Systems: Factory of the Future

Springer

Editors
Magdalena Hryb
Faculty of Mechanical Engineering
Poznań University of Technology
Poznań, Poland

Michał Wieczorowski
Faculty of Mechanical Engineering
Poznań University of Technology
Poznań, Poland

Piotr D. Moncarz
Center for Sustainable Development
and Global Competitiveness
Stanford University
Stanford, CA, USA

Jiehan Zhou
Shandong University of Science
and Techology
Qingdao, Shandong, China

Ewa Dostatni
Faculty of Mechanical Engineering
Poznań University of Technology
Poznań, Poland

ISSN 2195-4356 ISSN 2195-4364 (electronic)
Lecture Notes in Mechanical Engineering
ISBN 978-3-032-23037-9 ISBN 978-3-032-23038-6 (eBook)
https://doi.org/10.1007/978-3-032-23038-6

This Springer imprint is published by the registered company Springer Nature Switzerland AG
The registered company address is: Gewerbestrasse 11, 6330 Cham, Switzerland

Preface

This volume of Lecture Notes in Mechanical Engineering gathers accepted papers presented at the 9th International Scientific-Technical Conference MANUFACTURING 2026, held in Poznan, Poland, on May 19–21, 2026. The conference was hosted by the Faculty of Mechanical Engineering, Poznan University of Technology, Poland.

The primary objective of the MANUFACTURING 2026 was to showcase cutting-edge advancements across the broad spectrum of mechanical engineering, providing a dynamic forum for the exchange of innovative ideas and expert opinions. Focused on the theme of the Factory of the Future, the conference serves as a vital platform for dialogue on industrial evolution and science-industry synergy. Its global reach continues to expand, as evidenced by the consistently increasing participation from international research centers. Furthermore, the event fosters the integration of academia with the economic sector, encouraging collaboration with industry leaders and management professionals interested in implementing research results into business practice. The scope encompassed the following areas:

- Mechanical Engineering.
- Production Engineering.
- Quality Engineering.
- Measurement and Control Systems.
- Biomedical Engineering.

The global significance of the conference is reflected in the diversity of the submissions. The Organizing Committee received 219 contributions from 37 countries. To ensure the highest academic standards, all papers underwent a rigorous double-blind peer review process. Each manuscript was evaluated by at least three independent international experts, with a strict limit of five assignments per reviewer to guarantee depth and quality of feedback. Following this comprehensive appraisal, 120 papers were accepted for publication (an acceptance rate of 55%), representing the work of 347 authors from 25 countries. A total of 109 articles were published in the proceedings Advances in Manufacturing V, while 11 selected papers appeared in the journal Management and Production Engineering Review, issued under the auspices of the Committee on Production Engineering of the Polish Academy of Sciences and the Polish Society for Production Management.

The review process focused on originality of paper, significance of presented results, comprehensibility and quality of the presentation, relevance for this conference and overall impression. Furthermore, reviewers meticulously assessed the timeliness of the research, the adequacy of the literature review, methodological rigor, and the clarity of conclusions, as well as the quality of English. Feedback provided to authors included specific recommendations for improving methodology, data presentation, and editorial quality. Acceptance required at least two positive reviews and a full resolution of all reviewers' concerns. This international effort involved experts from 30 countries across Europe, North America, South Asia, and North Africa. Additionally, all manuscripts were checked using anti-plagiarism software, which allowed the similarity index to be less than 15%.

The book **Advances in Manufacturing V** is organized into five volumes that correspond to the main conference disciplines mentioned above.

The **Advances in Manufacturing V - Volume 4 - Measurement and Control Systems: Factory of the Future** volume focuses on contemporary challenges in metrology, advanced measurement techniques, and the digital transformation of quality control systems within the context of Industry 4.0 and 5.0. The collected contributions bridge fundamental research with practical industrial applications, covering both theoretical frameworks and real-world case studies. The content is organized into four thematic parts:

Metrology 4.0/5.0 – Concepts, Challenges, and Assessment Systems: Addresses the evolution of metrology toward a human-centered, data-driven paradigm, highlighting key technological and organizational challenges while focusing on the identification and parameterization of factors influencing coordinate measurement uncertainty and the use of database- and machine learning–based estimation methods.

Advanced Techniques and Methodologies for Surface and Geometry Measurement: Focuses on state-of-the-art optical and coordinate-based measurement methods, including repeatability assessment in focus variation microscopy, the influence of illumination conditions, resolution, and sampling density on measurement accuracy and repeatability, as well as laser tracker-based alignment and geometrical verification of industrial machinery and form deviation evaluation.

Analysis and Optimization of Manufacturing Processes and Tool Wear: Presents studies on surface topography, microstructure, and tool wear in machining of difficult-to-cut materials, the impact of tool geometry on surface defect removal, parameterization of abrasive belt grinding for complex geometries, and erosive and tribological performance analysis of conventionally and additively manufactured components.

Digitalization and Automation of Quality Control Systems in the Factory of the Future: Covers expert decision-support models for quality assessment, integration of digital inspection cards with QMS/MES systems in industrial case studies, and automation of geometric inspection processes through CAD data extraction, multi-criteria tool selection, and intelligent measurement planning.

Comprising 17 chapters prepared by 58 authors from 10 countries, this volume serves as a comprehensive reference for advancing modern metrology and quality engineering, fostering a strong synergy between advanced measurement science, digital technologies, and practical industrial applications.

We wish to express our sincere gratitude to the members of the Scientific and Program Committees for their dedication and expertise during the review process. We also extend our thanks to the authors, sponsors, and all contributors whose commitment was instrumental in making MANUFACTURING 2026 a resounding success.

May 2026

Magdalena Hryb
Jiehan Zhou
Michał Wieczorowski
Ewa Dostatni
Piotr D. Moncarz

Organization

Steering Committee

General Chair

Adam Hamrol	Poznan University of Technology, Poland

Chairs

Bartosz Gapiński	Poznan University of Technology, Poland
Magdalena Hryb	Poznan University of Technology, Poland
Justyna Trojanowska	Poznan University of Technology, Poland

Conference Founders

Zenobia Weiss (Deceased)	Poznan University of Technology, Poland
Adam Hamrol	Poznan University of Technology, Poland

Scientific Committee

Katarzyna Antosz	Rzeszow University of Technology, Poland
Christopher Brown	Worcester Polytechnic Institute, USA
Anna Burduk	Wrocław University of Science and Technology, Poland
Suthep Butdee	Rajamangala University of Technology Krungthep, Thailand
Marcin Butlewski	Poznan University of Technology, Poland
Olaf Ciszak	Poznan University of Technology, Poland
Reggie Davidrajuh	University of Stavanger, Norway
Ewa Dostatni	Poznan University of Technology, Poland
Jan Duda	Cracow University of Technology, Poland
Mosè Gallo	University of Naples Federico II, Italy
Bartosz Gapiński	Poznan University of Technology, Poland
Józef Gawlik	Cracow University of Technology, Poland
Adam Gąska	Cracow University of Technology, Poland
Arkadiusz Gola	Lublin University of Technology, Poland

Filip Górski	Poznan University of Technology, Poland
Marta Grabowska	Poznan University of Technology, Poland
Adam Hamrol	Poznan University of Technology, Poland
Magdalena Hryb	Poznan University of Technology, Poland
Jozef Husár	Technical University of Košice, Slovakia
Aminul Islam	Technical University of Denmark, Denmark
Vitalii Ivanov	Sumy State University, Ukraine
Michał Jakubowicz	Poznan University of Technology, Poland
Andrzej Jardzioch	West Pomeranian University of Technology in Szczecin, Poland
Sabahudin Jasarevic	University of Zenica, Bosnia & Herzegovina
Małgorzata Jasiulewicz-Kaczmarek	Poznan University of Technology, Poland
Anna Karwasz	Poznan University of Technology, Poland
Sławomir Kłos	Poznan University of Technology, Poland
Lucia Knapčíková	Technical University of Košice, Slovakia
Agnieszka Kujawińska	Poznan University of Technology, Poland
Vijaya Kumar Manupati	National Institute of Industrial Engineering, India
Stanisław Legutko	Poznan University of Technology, Poland
Sławomir Luściński	Kielce University of Technology, Poland
José Machado	University of Minho, Portugal
Damjan Maletic	University of Maribor, Slovenia
Dariusz Mikołajewski	Kazimierz Wielki University, Poland
Piotr Moncarz	Stanford University, USA
Magdalena Niemczewska-Wójcik	Cracow University of Technology, Poland
Przemysław Niewiadomski	University of Zielona Góra, Poland
Razvan Ioan Pacurar	Technical University of Cluj-Napoca, Romania
Piotr Paczos	Poznan University of Technology, Poland
Mirosław Pajor	Maritime University of Szczecin, Poland
Justyna Patalas-Maliszewska	University of Zielona Góra, Poland
Ivan Pavlenko	Sumy State University, Ukraine
Erwin Rauch	Free University of Bozen-Bolzano, Italy
Paulina Rewers	Poznan University of Technology, Poland
Izabela Rojek	Kazimierz Wielki University, Poland
Michał Rychlik	Poznan University of Technology, Poland
Krzysztof Santarek	Warsaw University of Technology, Poland
Sebastian Skoczypiec	Cracow University of Technology, Poland
Bożena Skołud	Silesian University of Technology, Poland
Jerzy Andrzej Sładek	Cracow University of Technology, Poland
Beata Starzyńska	Poznan University of Technology, Poland
Liangyan Tao	Nanjing University of Aeronautics and Astronautics, China

Anna Timofiejczuk	Silesian University of Technology, Poland
Justyna Trojanowska	Poznan University of Technology, Poland
Wiesław Urban	Bialystok University of Technology, Poland
Leonilde Varela	University of Minho, Portugal
Nikola Vitković	University of Niš, Serbia
Michał Wieczorowski	Poznan University of Technology, Poland
Ewa Więcek-Janka	Poznan University of Technology, Poland
Szymon Wojciechowski	Poznan University of Technology, Poland
Ralf Woll	BTU Cottbus-Senftenberg, Germany
Naiming Xie	Nanjing University of Aeronautics and Astronautics, China
Jiehan Zhou	Shandong University of Science and Technology, China

Program Committee

Adam Gąska	Cracow University of Technology, Poland
Adam Hamrol	Poznan University of Technology, Poland
Agnieszka Kaczmarek-Pawelska	University of Zielona Góra, Poland
Agnieszka Stachowiak	Poznan University of Technology, Poland
Alejandro Pereira	University of Vigo, Portugal
Aleksandar Miltenovic	University of Nis, Serbia
Alice Chlupová	Czech Academy of Sciences, Czech Republic
Alzbeta Sapietova	University of Žilina, Slovakia
Ancuta Pacurar	Technical University of Cluj-Napoca, Romania
André Guimarães	University of Beira Interior, Portugal
Andrzej Jardzioch	West Pomeranian University of Technology in Szczecin, Poland
Angelika Hofer	Free University of Bozen-Bolzano, Italy
Anna Borucka	Military University of Technology Warsaw, Poland
Anna Burduk	Wroclaw University of Science and Technology, Poland
Anna Karwasz	Poznan University of Technology, Poland
Anna M. Deptuła	Opole University of Technology, Poland
Anna Saniuk	University of Zielona Góra, Poland
Arkadiusz Gola	Lublin University of Technology, Poland
Bartosz Gapinski	Poznan University of Technology, Poland
Bogdan Dybała	Wroclaw University of Science and Technology, Poland
Christopher Brown	Worcester Polytechnic Institute, USA

Cristina Stefana Miron-Borzan	Technical University of Cluj-Napoca, Romania
Dagmara Łapczyńska	Wroclaw University of Science and Technology, Poland
Damian Grajewski	Poznan University of Technology, Poland
Damian Krenczyk	Silesian University of Technology, Poland
Dan-Sorin Comsa	Technical University of Cluj-Napoca, Romania
Daniel Medyński	Witelon Collegium State University, Poland
Dariusz Mikołajewski	Kazimierz Wielki University, Poland
Djahida Belayadi	National School of Cybersecurity, Algeria
Dražan Kozak	University of Slavonski Brod, Croatia
Edward Kozłowski	Lublin University of Technology, Poland
Emilia Campean	Technical University of Cluj-Napoca, Romania
Emilia Sabau	Technical University of Cluj-Napoca, Romania
Eryk Szwarc	Koszalin University of Technology, Poland
Ewa Dostatni	Poznan University of Technology, Poland
Ewa Więcek-Janka	Poznan University of Technology, Poland
Gabriel Stolárik	Technical University of Košice, Slovakia
Grzegorz Pająk	University of Zielona Góra, Poland
Hanna Łosyk	University of Zielona Góra, Poland
Hubert Wojciechowski	Poznan University of Technology, Poland
Izabela Kudelska	Poznan University of Technology, Poland
Izabela Rojek	Kazimierz Wielki University, Poland
Jacek Diakun	Poznan University of Technology, Poland
Jakub Kaščák	Technical University of Košice, Slovakia
Jakub Kopowski	Kazimierz Wielki University, Poland
Jerzy Józwik	Lublin University of Technology, Poland
Joanna Cyganiuk	University of Zielona Góra, Poland
Joanna Helman	Wroclaw University of Science and Technology, Poland
Joanna Wilczarska	Bydgoszcz University of Science and Technology, Poland
Jozef Husár	Technical University of Košice, Slovakia
Jozef Török	Technical University of Košice, Slovakia
Józef Kuczmaszewski	Lublin University of Technology, Poland
Justyna Patalas-Maliszewska	University of Zielona Góra, Poland
Kamil Krot	Wroclaw University of Science and Technology, Poland
Kamil Leksycki	University of Zielona Góra, Poland
Kamil Musiał	Wroclaw University of Science and Technology, Poland
Karol Dąbrowski	University of Zielona Góra, Poland
Katarzyna Antosz	Rzeszow University of Technology, Poland

Katarzyna Cheba	West Pomeranian University of Technology in Szczecin, Poland
Katarzyna Kazimierska-Drobny	Kazimierz Wielki University, Poland
Katarzyna Skrzypek	University of Zielona Góra, Poland
Khrystyna Berladir	Sumy State University, Ukraine
Konrad Kobiela	Cracow University of Technology, Poland
Krzysztof Kalinowski	Silesian University of Technology, Poland
Krzysztof Nowacki	Silesian University of Technology, Poland
Krzysztof Żywicki	Poznan University of Technology, Poland
Leonilde Varela	University of Minho, Portugal
Lucia Knapčíková	Technical University of Košice, Slovakia
Łukasz Rymaniak	Poznan University of Technology, Poland
Magdalena Hryb	Poznan University of Technology, Poland
Magdalena Rybaczewska-Błażejowska	Kielce University of Technology, Poland
Magdalena Żukowska	Poznan University of Technology, Poland
Małgorzata Śliwa	University of Zielona Góra, Poland
Marcin Suszynski	Poznan University of Technology, Poland
Marek Kočiško	Technical University of Košice, Slovakia
Marek Macko	Kazimierz Wielki University, Poland
Maria Rosienkiewicz	Wroclaw University of Science and Technology, Poland
Marian Bartoszuk	Opole University of Technology, Poland
Marius Sorin Pustan	Technical University of Cluj-Napoca, Romania
Marks Gorohovs	Riga Technical University, Latvia
Marta Grabowska	Poznan University of Technology, Poland
Marta Harničárová	Slovak University of Agriculture in Nitra, Slovakia
Michał Jakubowicz	Poznan University of Technology, Poland
Michał Rogalewicz	Poznan University of Technology, Poland
Mihai Dragomir	Technical University of Cluj-Napoca, Romania
Milan Edl	University of West Bohemia, Czech Republic
Mircea Fulea	Technical University of Cluj-Napoca, Romania
Mukund Harugade	Padmabhooshan Vasantraodada Patil Institute of Technology, India
Natalya Shramenko	Lviv Polytechnic National University, Ukraine
Nataša Náprstková	University of Jan Evangelista in Ústí nad Labem, Czech Republic
Nguyen Hong Vi	Vietnamese German University, Vietnam
Nikola Vitković	University of Nis, Serbia
Olaf Ciszak	Poznan University of Technology, Poland
Olha Kalman	Technical University of Košice, Slovakia

Paulina Rewers	Poznan University of Technology, Poland
Paweł Buń	Poznan University of Technology, Poland
Paweł Szczygieł	Kielce University of Technology, Poland
Paweł Zmarzły	Kielce University of Technology, Poland
Pedro Torres	Polytechnic Institute of Castelo Branco, Portugal
Piotr Gąska	AGH University of Krakow, Poland
Piotr Kotlarz	Kazimierz Wielki University, Poland
Piotr Trojanowski	West Pomeranian University of Technology in Szczecin, Poland
Piotr Ziembicki	University of Zielona Góra, Poland
Przemysław Zawadzki	Poznan University of Technology, Poland
Radosław Wichniarek	Poznan University of Technology, Poland
Răzvan Păcurar	Technical University of Cluj-Napoca, Romania
Roman Kielec	University of Zielona Góra, Poland
Sandra Grabowska	Silesian University of Technology, Poland
Sebastian Saniuk	University of Zielona Góra, Poland
Sebastian Skoczypiec	Cracow University of Technology, Poland
Sergej Hloch	Technical University of Košice, Slovakia
Sławomir Kłos	University of Zielona Góra, Poland
Stanislaw Legutko	Poznan University of Technology, Poland
Stella Hrehova	Technical University of Košice, Slovakia
Sven Maricic	Juraj Dobrila University of Pula, Croatia
Sylvia Kusmierczak	University of Jan Evangelista in Ústí nad Labem, Czech Republic
Szymon Wojciechowski	Poznan University of Technology, Poland
Thanh Tran	RMIT University, Vietnam
Tomasz Bednarek	Kazimierz Wielki University, Poland
Vijaya Kumar Manupati	NITIE Mumbai, India
Vitalii Ivanov	Sumy State University, Ukraine
Wiesław Kuczko	Poznan University of Technology, Poland
Wiktor Szot	Kielce University of Technology, Poland
Wojciech Babirecki	University of Zielona Góra, Poland
Wojciech Macek	Gdańsk University of Technology, Poland
Yuliia Denysenko	Sumy State University, Ukraine
Zbigniew Wiśniewski	Lodz University of Technology, Poland
Zuzana Mitaľová	Technical University of Košice, Slovakia

Special Sessions and Special Session Organizing Committee

Efficiency-Driven Manufacturing: Tools and Trends in the Era of Industry 4.0

Justyna Trojanowska	Poznan University of Technology, Poland
Leonilde Varela	University of Minho, Portugal
Vijaya Kumar Manupati	NITIE Mumbai, India
Thanh Tran	RMIT University, Vietnam
Jozef Husár	Technical University of Košice, Slovakia

Intersecting Quality Management and Mechanical Design for Manufacturing Sustainability

Florin Popişter	Technical University of Cluj-Napoca, Romania
Diana Cristina Dragomir	Technical University of Cluj-Napoca, Romania
Corina Julieta Birleanu	Technical University of Cluj-Napoca, Romania
Mircea Claudiu Fulea	Technical University of Cluj-Napoca, Romania

Human Factors in Manufacturing

Marcin Butlewski	Poznan University of Technology, Poland
Beata Mrugalska	Poznan University of Technology, Poland
Wiktoria Czernecka	Poznan University of Technology, Poland
Fryderyk Wachowiak	Poznan University of Technology, Poland
Krzysztof Hankiewicz	Poznan University of Technology, Poland

Circularity in Advanced Production Systems: Materials, Methods, and Digital Transformation

Marek Kočisko	Technical University of Košice, Slovakia
Lucia Knapčikova	Technical University of Košice, Slovakia
Jakub Kaščak	Technical University of Košice, Slovakia
Jozef Török	Technical University of Košice, Slovakia

Smart Manufacturing and Sustainability: Interdisciplinary Approaches in the Era of Industry 4.0/5.0

Radosław Maruda	University of Zielona Góra, Poland
Sebastian Saniuk	University of Zielona Góra, Poland
Katarzyna Skrzypek	University of Zielona Góra, Poland
Daria Moskwa - Bęczkowska	Kielce University of Technology, Poland
Karol Dąbrowski	University of Zielona Góra, Poland

From Additive Manufacturing to Healthcare Innovation through Digital Transformation in Biomedical Engineering

Filip Górski	Poznan University of Technology, Poland
Răzvan Păcurar	Technical University of Cluj-Napoca, Romania
Nikola Vitkovic	University of Niš, Serbia
Michał Rychlik	Poznan University of Technology, Poland

Intelligent Adaptive Systems in Integrated Production Management Processes

Mariusz Piechowski	Merito WSB University in Poznan, Poland
Małgorzata Jasiulewicz-Kaczmarek	Poznan University of Technology, Poland
Katarzyna Antosz	Rzeszow University of Technology, Poland
Ryszard Wyczółkowski	Silesian University of Technology, Poland
Arkadiusz Gola	Lublin University of Technology, Poland

Intelligent Solutions for Industry of the Future

Izabela Rojek	Kazimierz Wielki University in Bydgoszcz, Poland
Ewa Dostatni	Poznan University of Technology, Poland
Dariusz Mikołajewski	Kazimierz Wielki University in Bydgoszcz, Poland
Nataša Náprstková	University of Jan Evangelista in Ústí nad Labem, Czech Republic

Grey Systems Modeling for Engineering Solutions and Business Optimization

Magdalena Hryb	Poznan University of Technology, Poland
Agnieszka Kujawińska	Poznan University of Technology, Poland
Naiming Xie	Nanjing University of Aeronautics and Astronautics, China
Liangyan Tao	Nanjing University of Aeronautics and Astronautics, China
Rafał Mierzwiak	Poznan University of Technology, Poland

Intelligent Methods Supporting Manufacturing Systems Efficiency

Anna Burduk	Wroclaw University of Science and Technology, Poland
Damian Krenczyk	Silesian University of Technology, Poland
Dagmara Łapczyńska	Wroclaw University of Science and Technology, Poland
Kamil Krot	Wroclaw University of Science and Technology, Poland

Contents

Analysis and Optimization of Manufacturing Processes and Tool Wear

Digitalization and Automation of Quality Control Systems in the Factory of the Future

Metrology 4.0/5.0 – Concepts, Challenges, and Assessment Systems

Determinants and Challenges of Metrology 5.0

Michał Wieczorowski[1](✉), Andrzej Kurkiewicz[2], Maxence Bigerelle[3], Alejandro Pereira[4], Grzegorz Królczyk[5], Francois Berkmans[3], and Bartosz Gapiński[1]

[1] Faculty of Mechanical Engineering, Poznan University of Technology, Poznan, Poland
michal.wieczorowski@put.poznan.pl
[2] Institute of Public Affairs, Jagiellonian University in Krakow, Krakow, Poland
[3] Polytechnic University of Hauts-de-France, LAMIH, Valenciennes, France
[4] University of Vigo, Manufacturing Engineering Group, Vigo, Spain
[5] Independent Researcher, Opole, Poland

Abstract. The article presents the concept of modern metrology as a response to the challenges of industry and digital transformation. It shows that Metrology 5.0 represents the next stage in the development of measurement science, combining issues of accuracy with intelligent technologies and a human-centered approach. The human being becomes the central subject of the industrial revolution, not only supervising processes but also co-creating solutions based on data and artificial intelligence. Attention is drawn to the problems of the generation entering the labor market and the way it perceives the surrounding world. The evolution of industry and metrology from version 1.0 to 5.0 is discussed, emphasizing the importance of measurements for production quality and innovation. On the path toward Metrology 5.0, a key role is played by digital technologies, artificial intelligence, the Internet of Things, digital twins, and multisensor measurement systems that support real-time data analysis. The article also points out technological challenges such as cybersecurity, data reliability, system integration, and the need to comply with the principles of sustainable development. Metrology 5.0 is therefore not only a technological innovation but also a social and cultural change, requiring a new approach to education, ethics, and human–technology collaboration.

Keywords: Metrology 5.0 · Digital Twin · Artificial Intelligence · Measurement Systems · Human–Machine Collaboration

1 Introduction

Metrology, defined as the science of measurement - its principles, methods, technical means, and the interpretation of results - is one of the oldest and most comprehensive fields of technical knowledge. It also constitutes one of the fundamental pillars of technological civilization. It is almost a truism to say that measurements are indispensable in every aspect of human activity, from science and various branches of industry to safety and environmental protection. Since the dawn of humankind, technological progress has

M. Hryb et al. (Eds.): MANUFACTURING 2026, LNME, pp. 3–16, 2026.
https://doi.org/10.1007/978-3-032-23038-6_1

been inseparably linked with the development of measurement methods. Every innovation or even technological improvement - that is, any implemented change - requires an accurate description of physical reality, which is possible only through measurement, by assigning a specific value to a given quantity. The reliability of these measurements is ensured by a global metrological infrastructure developed across countries worldwide, thanks to which results obtained in different laboratories and nations are comparable and traceable. This enables the assurance of production quality and the reliable certification of products. As technological progress continues, metrology is transforming from a purely technical discipline into a key element of the knowledge infrastructure, allowing trustworthy communication between the physical and digital worlds [1].

The dynamic development of measurement technologies, cyber-physical systems, and artificial intelligence is leading to a redefinition of the role of metrology in both the economy and science [2]. Throughout successive industrial revolutions, the role of this science has evolved from a simple tool for quality control to a crucial component of data ecosystems. Today, we are witnessing the emergence of the Metrology 5.0 concept, corresponding to the needs of the Industry 5.0 era - a world focused not only on automation and efficiency but also on collaboration between humans and intelligent technologies, as well as on sustainable development. The concept of Metrology 5.0 thus represents, in a sense, a fusion of measurement precision, intelligent production and management systems, and sustainability.

The fundamental function of metrology in the economy is to ensure measurement traceability, that is, the linkage of all measurements to national and international standards of units (Fig. 1).

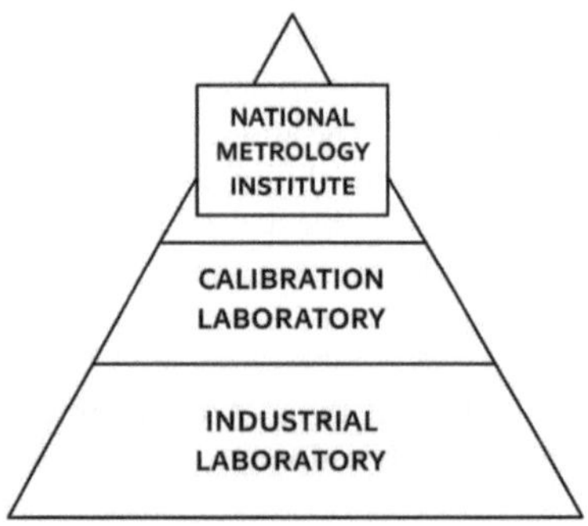

Fig. 1. Measurement traceability in the national context.

Thanks to this, it becomes possible to compare measurement results regardless of the place and time of their execution. Every economic decision is based on measurement data, whose uncertainty translates into financial, legal, and technological risk. Therefore, a well-developed metrological system constitutes an element of a country's critical infrastructure. Metrology is also one of the pillars of the knowledge-based economy, enabling technological development and the transfer of innovation. Modern technologies cannot exist without accurate measurements of various quantities. In turn, these measurements drive technological progress, stimulating the development of new industries and high value-added products. Furthermore, metrology contributes to improving the competitiveness of the economy through appropriate quality control, while also enabling the optimization of production processes, reduction of waste, control of energy and raw material consumption, and, consequently, increasing economic efficiency. At the macroeconomic level, metrology ensures the reliability of trade and the stability of commercial relations. It is essential to guarantee that units of measurement (e.g., kilogram, liter, kilowatt-hour) have the same meaning in all countries. Thanks to the establishment of common metrological standards, enterprises from different nations can cooperate without the need to repeat testing and certification, which reduces transaction costs and accelerates the introduction of products to the market. However, the importance of metrology extends far beyond the economic sphere, encompassing social and environmental aspects as well. Verification of outcomes through measurements is indispensable in healthcare (e.g., medical diagnostics, pharmaceutical production), in monitoring the quality of air, water, and soil, and in energy and transportation sectors, where metrology enables the assessment of sustainable development policies. In addition, metrology plays a crucial role in national security, supporting the functioning of the defense industry, nuclear energy, and technical safety systems. Accurate measurements form the basis for the technical safety of critical infrastructure as well as for emergency response, logistics, and information systems. It is in these areas that one can most clearly see how metrology is becoming a science of trust in data, which serve as the fuel of the 21st-century economy.

Although metrology has traditionally been associated with physics and engineering, today it represents a distinctly interdisciplinary science, combining achievements from many fields - from natural and technical sciences, through computer science, to economics, law, management, and the social sciences. The interdisciplinary nature of metrology arises from the fact that measurement is the fundamental tool for understanding reality in almost every area of science and technology. Metrology is therefore the universal language of the empirical sciences, ensuring the comparability and reliability of research results regardless of the field. Technological progress in recent decades has deepened the interdisciplinary character of metrology. An important aspect of this evolution is its growing influence on the social sciences and the economy. The system of weights and measures, and the assurance of measurement uniformity in trade, medicine, or environmental protection, form the foundation of social trust and a basic condition for the functioning of modern societies. Metrology thus supports decision-making processes in management, economics, and technology policy, building a bridge between science, industry, and public administration. Contemporary metrology encompasses three fundamental levels: scientific metrology, which develops theoretical foundations and maintains standards of measurement units; industrial metrology, concerned with ensuring measurement traceability in production and technology and legal metrology, which guarantees compliance of measurements with regulations and the protection of public interests. Each of these levels requires the collaboration of multiple scientific disciplines, reflecting the comprehensive and integrative nature of modern metrology.

With the development of the digital industry and the data-driven economy, the importance of metrology as a science ensuring the quality of information is also increasing.

Measurement data constitute not only the foundation for the operation of measurement systems (including their automation) but are also indispensable for artificial intelligence, process prediction, modeling, and similar applications. If the input data are burdened with excessive error, if the uncertainty in the measurement process is significant, or if there is no metrological traceability, even the most advanced analytical algorithms will lead to incorrect conclusions [3, 4]. From this point of view, metrology - formerly associated mainly with instruments, their calibration, and the measurement process - is now becoming a science of data reliability in the digital environment. One may venture to say that modern metrology is undergoing a transformation from a science of accuracy to a science of trust. In the digital era, what becomes increasingly important is not only how much we measure, but how we ensure the consistency and traceability of data within complex information systems.

Modern measurement technologies are becoming increasingly and more deeply integrated with production systems within the frameworks of Industry 4.0 and 5.0, resulting in the creation of intelligent, networked measurement environments. Intensive data acquisition from industrial processes is made possible, among other factors, by the miniaturization of sensors, the advancement of electronics, and the integration of systems, for example within the Internet of Things (IoT) architecture. At the same time, there is growing discussion about the introduction of Digital Twins to virtually replicate measurement systems, opening up new possibilities for process verification, uncertainty analysis, calibration monitoring, and simulation-based prediction of results [5]. In this paper, therefore, we aim to indicate the directions of metrology development in the context of Industry 5.0 and the emerging new paradigms of science and technology.

2 The Human Being as the Subject of the Industrial Revolution

Successive social and industrial revolutions are becoming increasingly oriented toward serving human needs. In recent years, it has become evident that our society operates in a world where the boundary between physical and digital reality is gradually disappearing. This is particularly true for the younger generation entering the labor market, which has grown up in an environment characterized by an unprecedented pace of social, technological, cultural, and economic change [6]. For people born after the year 2000, change has ceased to be an event or even an exception - it has become an everyday occurrence and, in a sense, the natural state of the world. This generation is coming of age in an era of digital transformation, global challenges, and - unfortunately - growing socio-economic uncertainty, which to a large extent shapes its worldview, values, and lifestyle.

The most evident change that young people observe around them is the digitalization of nearly all spheres of life [7, 8]. The Internet, social media, artificial intelligence, online life, and widespread streaming have become fundamental tools for learning, communication, and entertainment. Observing the behavior of this generation, one can see how traditional forms of interaction, work, and education are giving way to virtual relationships, remote work, and online learning [9]. At the same time, young people - at least to some extent - are increasingly aware of the dual nature of these changes, which on one hand provide virtually unlimited access to knowledge, career opportunities, and global collaboration, but on the other lead to concerning phenomena such as technology addiction, disinformation (often in a subtle or disguised form), and the superficiality of interpersonal relationships. As a result, psychologists increasingly call for digital balance and conscious use of technology. However, this is not the whole picture. The digital revolution overlaps with deep social and cultural transformations. Social norms and customs are changing, and the importance of gender equality, tolerance, human rights, and cultural diversity is growing. The generation entering the labor market is also more open to issues of identity, ecology, and social justice, while being less attached to traditional structures - such as the hierarchical family model. There is also a rising awareness of global interconnectedness, and - despite certain countertrends - of the world as a space without rigid borders restricting the movement of people and goods. Young people recognize that their lives are part of a complex global system, in which political, economic, and environmental decisions made in one region of the world affect the situation in another [10]. This awareness fosters the development of prosocial attitudes, volunteerism, and civic engagement, especially in areas such as environmental protection, social equality, and education. For some, ecological awareness and responsibility for the planet have even become elements of personal identity and moral values, expressed through a lifestyle based on sustainable development, reduced consumption, a positive attitude toward recycling, and the use of public transport and renewable energy sources. The concept of life success is also changing - career is no longer seen as the ultimate life goal. Intangible values such as self-realization, psychological balance, a sense of purpose, and the quality of relationships are gaining importance. In professional life, flexibility, job mobility, and remote work are becoming increasingly valued. At the same time, mental health awareness is growing. Topics such as stress, depression, and burnout are no longer taboo, and open discussions about emotions have become the norm. This marks one of the most significant cultural shifts of recent years.

It is also important to note that the generation growing up in the 21st century has experienced three major global crises [11], each of a very different nature: the financial crisis of 2008, the COVID-19 pandemic, and the war in Ukraine. Each of these events has shaped the worldview, value system, and approach to life, work, and social relationships of young people in different ways. The 2008 financial crisis undermined the sense of stability in the financial system and the security of individual savings, primarily affecting the so-called millennial generation, who were entering adulthood during the recession. Many began their professional careers amid economic uncertainty, delaying decisions such as starting a family or buying a home. The collapse of banks and financial markets eroded trust in institutions and capitalist structures, encouraging the search for alternative forms of economic security, such as the sharing economy or investments in cryptocurrencies. This crisis was also the first signal of a shift in value hierarchies, as flexibility, independence, and personal development began to take precedence over the once-coveted notion of a "stable lifelong career." A dozen years later, the COVID-19 pandemic struck Generation Z, who, during their key educational years and the transition to adulthood, experienced social isolation. School closures and remote learning led to a deterioration in mental health, disrupting the period of forming friendships and peer relationships [12, 13]. Rates of depression and anxiety increased, and interpersonal interactions moved into the digital sphere. On one hand, the pandemic accelerated the digitalization of life, making young people the most "online" generation in history, but on the other, it made them the most vulnerable to social media addiction and the pressures of virtual life. In the professional sphere, there was a further reassessment of attitudes toward work, with a growing pursuit of flexible employment models, meaningful careers, and work-life balance. The next crisis, the war in Ukraine, brought a renewed sense of insecurity and uncertainty about the future. An armed conflict after decades of peace in Europe came as a profound shock. It heightened feelings of existential fear and awareness of the power of disinformation. The war and its economic consequences - such as inflation and rising energy prices - had a particularly strong impact on those just beginning independent lives. All these experiences have made today's young generation grow up in a world of chronic uncertainty. Crises and instability have become a defining generational experience, giving rise to an ambivalent attitude: on one hand, fear of the future and a sense of lack of control over its course, and on the other, motivation to act, empathy, social responsibility, and the search for sustainable values. Each crisis has also accelerated the process of digital transformation, making this generation the most technologically proficient - but also the most psychologically burdened - in history. As a result, a society has emerged that is learning to live in a state of permanent crisis - flexible, aware, yet anxious about the future. This generation does not merely observe change; it actively participates in shaping it, redefining the concepts of work, success, relationships, and community in the spirit of social responsibility, innovation, and adaptability, expressing the conviction that if the world changes, we must change with it.

Metrology 5.0 also brings about a shift in the competencies of the classical metrologist. The act of measurement itself is ceasing to be the domain of the operator; however, the expectations placed on such professionals by management will change significantly in the coming years. Although automation remains one of the pillars of modern metrology, the human element remains indispensable and will continue to occupy a central role in the measurement process (human-in-the-loop). The metrologist of the future will need to understand not only the instruments and their interaction with the measured object but also, to some extent, mathematical models and decision-making systems (both control and supervisory), which will be deeply integrated into measurement processes. Thus, the role of the metrologist is shifting from operational to analytical and interpretive, encompassing tasks such as AI algorithm validation and the assessment of their accuracy and reliability within real-world processes. This evolution naturally requires new competencies: data analysis and processing principles, basic programming skills, knowledge of AI algorithms, statistics from the perspective of artificial intelligence, and - most importantly - cybersecurity awareness. A cyberattack that distorts or alters data, changing correct measurements into incorrect ones (or vice versa), can have critical consequences, not only for the functioning of the employing organization but also for its clients and end users.

3 Industry and Metrology

Metrology, as a scientific discipline, plays a key role in the development and operation of industry. It constitutes not only the technical foundation for quality control, but also a strategic element in the management of production processes, safety, innovation, and competitiveness. Modern industry, increasingly based on automation and digitalization, must rely on feedback systems built on consistent and reliable measurements. In this context, metrology emerges as a source of trust in measurement results throughout the entire value chain - from research and development to the final product [14, 15].

Since the very beginning of industrial development, measurements have been the foundation of product quality assurance. With the advent of mass production, the need arose to standardize dimensions, tolerances, and measurement norms, leading to the establishment of national and international metrological institutions. Modern industrial enterprises use measurement systems to verify product compliance with standards and customer requirements, optimize technological processes, detect and eliminate deviations in real time, and document quality within management systems. In this sense, metrology serves as the technical language of industry, ensuring the consistency of requirements among manufacturers, suppliers, and customers.

Every country that takes its economy seriously ensures proper collaboration between industry and the national metrological infrastructure. To this end, it maintains a system of institutions that guarantee measurement traceability, ranging from National Metrology Institutes (NMI) to accredited calibration and testing laboratories. In Poland, this function is performed by the Central Office of Measures (GUM), which cooperates with industry in the areas of calibration, certification, and the development of measurement standards. Through this collaboration, industry can be confident that its measurements are reliable, and customers can trust that they receive accurately manufactured products. In this context, a number of new initiatives have naturally emerged in response to the digital revolution - for example, the Digital Calibration Certificate (DCC), which enables digital exchange of calibration data, thereby accelerating information flow while enhancing the security of measurement data.

In line with the principles of Industry 5.0 and the United Nations Sustainable Development Strategy, metrology plays a vital role in monitoring energy efficiency, greenhouse gas emissions, air quality, and resource utilization. Through appropriate measurements, it becomes possible to assess the environmental impact of technologies, making metrology not only a technical tool but also an ethical and social instrument that supports responsible industrial decision-making [16, 17]. Integrating metrological activities with the concept of Sustainable Development also contributes to optimizing measurement processes in terms of energy and material consumption. The importance of metrology in industry has not only a technical but also a significant economic dimension. Investments in metrological systems directly translate into increased efficiency, reduction of material losses, and improvement in product quality. Such investments also yield a high economic return, estimated at 3–5 units of added value for every unit invested. Moreover, metrology helps reduce the costs of errors and product complaints, increase productivity, support innovation, shorten time-to-market, and enhance the competitiveness of enterprises. By ensuring traceability - that is, linking measurement results to SI units - metrology enables transparent data exchange within global supply chains [18, 19].

4 On the Road to Metrology 5.0

With each successive stage of the industrial revolutions, from mechanization to the integration of cyber-physical systems, the role of metrology has evolved - from a simple quality control tool to a key component of complex data ecosystems [20]. However, metrology's importance can be traced back to ancient Egypt, where pyramid builders used length standards (most often made of wood) that were periodically compared to stone standards, considered more reliable. Throughout history, the development of metrology has been closely correlated with subsequent industrial revolutions. The First Industrial Revolution brought about the era of mechanical instruments and manual measurements (including the first vernier calipers). Metrology 2.0 marked the next step—awareness of standardization and the need to ensure measurement accuracy. It was during this period that reference standards such as the meter - originally defined as the distance between two engraved lines - were established. In the second half of the 20th century, metrology experienced a renaissance with the introduction of the coordinate system, giving rise to coordinate measuring techniques (CMM). This allowed specific dimensions to be calculated based on the coordinates of points collected from geometric shapes and solids, marking the beginning of measurement automation as well as data recording and analysis systems. Metrology 4.0, emerging in the second decade of the 21st century, is the direct metrological counterpart of Industry 4.0 principles. It encompasses big data, the development of optical scanning techniques, digitalization, freeform surfaces and point cloud analysis, metrology for additive manufacturing (3D printing) [21], multisensor systems, sensor networks, the Internet of Things (IoT), communication between measurement and production systems, real-time data analysis, and system interoperability [22, 23]. The growth of Industry 4.0 also introduced the concept of Smart Metrology - intelligent measurement systems as an integral part of the production process. Modern sensors, optical systems, coordinate measuring systems at various scales [24], computed tomography (CT) [25], and laser scanning systems [26] can now be directly integrated into production lines and connected with control systems, forming in-line measurement systems that enable active process control and real-time correction [27]. 3D printing has also become widespread, bringing with it a new metrological approach that takes into account the specific aspects of various additive manufacturing technologies [28].

As already mentioned, Metrology 5.0 responds to the needs of a world focused on human collaboration with intelligent technologies. For this reason, analyzing the behaviors and needs of the younger generation - as today's and tomorrow's workforce - is particularly important. We are reaching a stage where manufacturing and measurement systems are designed to cooperate with humans, utilizing artificial intelligence, digital twins, and the principles of sustainable development. Among the defining features of Industry 5.0 is its human-centric approach, which places the human being at the center of the industrial process [29, 30]. What matters most is employee well-being, creativity, and job satisfaction, while technology is intended to support rather than replace humans. A key element of this paradigm is the collaboration between humans and robots (including cobots - collaborative robots), which are designed to assist humans in repetitive and physically demanding tasks, leaving creative and strategic work to people. There is also a renewed emphasis on personalized production, through which goods and services are increasingly tailored to individual customer needs by leveraging artificial intelligence, big data, and the Internet of Things. However, this mass personalization carries certain limitations, as it is well known that mass production remains more cost-effective than individual manufacturing. By focusing on the human element, Industry 5.0 also incorporates sustainability and ecological responsibility, emphasizing energy-efficient processes, the circular economy, waste minimization, and environmentally friendly technologies. Ethics and accountability play an equally important role, particularly concerning privacy protection, respect for workers' rights, and the responsible use of AI. Effective collaboration between humans, robots, and artificial intelligence in the coming years is therefore viewed as crucial to the industry of the future [31]. Another significant factor is the impact of recent global crises and the resulting disruptions in supply chains, which have shifted industrial priorities toward resilience and flexibility - systems must now be capable of rapidly adapting to changing economic environments. All these elements are reflected in metrology as well. The key difference between Metrology 4.0 and 5.0 lies in the shift of paradigm - from full automation and the elimination of humans to a symbiosis between humans and machines, redefining the relationship between people, technology, and data, with an emphasis on collaboration, transparency, and responsibility [32]. Measurements no longer serve solely to monitor quality, but also to support adaptive production processes in which humans and intelligent systems cooperate in real time. An example of this is integrated measurement and diagnostic systems that use artificial intelligence to predict errors and automatically compensate for deviations [33].

5 Further Technological Challenges

When considering the future of metrology in the modern world, it is worth analyzing the key technologies driving Metrology 5.0, at least in the form we know it today. At the forefront stands artificial intelligence (AI) and its subset focused on machine learning, which enables systems to learn from data without the need to explicitly program every rule. It can be stated with strong conviction that the application of machine learning algorithms in metrology will, in the future, allow for even more advanced automatic anomaly detection, error classification, and measurement uncertainty prediction. Neural network-based models will learn patterns from measurement data to support the decision-making processes of metrologists, such as the selection of measurement methods or calibration strategies. Artificial intelligence will not only assist but also adaptively adjust measurement parameters in real time, becoming an integral component of information-rich metrology [34]. Another technology that will play an increasingly important role is the concept of Digital Twins [35, 36]. In metrology, they integrate measurement, data analysis, and real-time simulation by creating virtual replicas of physical objects, devices, or measurement processes. These virtual models are continuously updated based on sensor data within the Internet of Things (IoT) environment. As a result, the digital model faithfully reflects the state and behavior of the physical system, enabling better quality control, process optimization, and prediction of potential deviations or failures. In this way, metrology becomes an integral part of the digital quality loop, linking measurement, data analysis, and control decisions. It becomes possible to simulate the influence of environmental factors (such as temperature or vibration) and component wear on measurement results, assess measurement uncertainty in real time, and even optimize measurement sequences without interfering with the physical system. In laboratory metrology, Digital Twins will form the foundation for hybrid laboratories that combine physical and virtual measurements [37]. One of the most important applications of Digital Twins in metrology will be virtual calibration and prediction of measurement errors, which will ultimately reduce the number of physical calibrations required - lowering maintenance costs and minimizing downtime. Another application is the simulation of measurement processes, allowing engineers to test various strategies in a virtual environment - optimizing probe paths, selecting styli, tips, or lenses, and verifying measurement time - before programming and performing the actual inspection of the measured component. Digital Twins will also play a major role in data management within quality assurance systems. Integrated with other production-support systems, they will enable real-time quality monitoring, trend analysis, and support for statistical process control (SPC). Data generated by digital models will also contribute to predictive maintenance, forecasting instrument failures or the need for recalibration based on sensor data. Another key technology that will see significant growth is the Internet of Things (IoT). It can be assumed that smart sensors - analyzing the condition of systems, devices, subsystems, and their components - will be embedded virtually everywhere. IoT networks will interconnect thousands of sensors across distributed industrial environments. This continuous real-time monitoring will provide instant information about verification, regulation, and control actions taken by the system, and, when necessary, prompt the operator to perform specific tasks. This will be supported by cloud computing technologies, enabling centralized management and data analysis from multiple sources, based on cyber-physical systems. These developments will lead to intelligent production control with automatic parameter correction, resulting in self-adaptive measurement systems capable of evaluating their own errors and measurement uncertainty, and performing self-calibration according to defined measurement procedures [38]. This represents a new level of quality control and assurance, as well as a new standard of trust in metrological data. In this context, cybersecurity becomes essential. Moreover, blockchain technology may further enhance data integrity, providing immutability and full traceability of measurement records - particularly important in areas where measurement data form the basis for critical decisions. This will also be a key issue in validating results across distributed industrial environments and in ensuring transparency and ethical use of measurement data.

Among the technologies described above, several potential risks can also be identified. One of them - already mentioned - is cybersecurity in measurement systems and data, both in laboratory and industrial environments. This issue is becoming critical due to the growing number of network connections, which increases vulnerability to cyberattacks and data manipulation. Many organizations already perceive this as such a serious threat that their production systems are isolated from external networks for safety reasons. Despite significant technological advances, the implementation of the Metrology 5.0 concept still faces numerous challenges and limitations. One of them is the lack of standardized data formats and poor system interoperability, which makes it difficult to exchange measurement data between different systems. Although initiatives have been undertaken by the International Organization for Standardization (ISO) to develop such formats, in many areas the situation remains far from ideal. Another challenge involves the integration of legacy systems with new technologies, since it is neither feasible nor economical to replace all existing devices and systems with entirely new ones in a short time. Nonetheless, Metrology 5.0 undoubtedly requires investment in infrastructure and personnel training. A further aspect concerns artificial intelligence itself. While AI represents one of the most promising elements of modern technology, it still suffers from the problem of hallucinations - the tendency of models to produce inaccurate or fabricated information. At present, there are no effective methods for assessing the trustworthiness of AI-based models, which limits confidence in results generated with their assistance. Moreover, when results are produced by machine learning or algorithmic systems, there are no established procedures for determining measurement uncertainty associated with them. This raises a significant ethical and legal issue, encapsulated in the question: Who bears responsibility for erroneous decisions based on data processed by algorithms? Addressing this will require new regulations and legal frameworks, similar to those being developed for autonomous vehicles and the accidents they may cause.

These challenges alone demonstrate that technological progress must go hand in hand with the development of methodologies, standards, and human competencies. To achieve this, pilot hybrid laboratories should be established in the coming years, where real measurements are complemented by simulations in digital environments. Such integration of the physical and digital worlds - including through the use of augmented and virtual reality - will later be transferred into industrial practice. Equally important will be the adaptation of metrological education to the needs of the economic and business environment, or even the creation of a new generation of metrological education, combining engineering knowledge with digital and ethical competencies [39]. In the same direction, doctoral programs should evolve, focusing on research and the practical implementation of developed solutions [40, 41].

6 Conclusions

Metrology 5.0 is not only the future of laboratories but - perhaps even more importantly - the direction of development for the entire data-driven economy. The path toward it is no longer solely about precision but about developing a smart, responsible, and integrated science of measurement for the future. The whole idea represents a new paradigm in measurement science - one that combines accuracy, intelligence, and responsibility. Its essence lies not merely in automation, but in the harmonious collaboration between humans and technology, based on trusted data, artificial intelligence, and the principles of sustainable development. Achieving full implementation of this concept will require the establishment of methodological, legal, and educational frameworks that enable the creation of a science that is not only technological but also responsible and ethical. This synergy of knowledge, technology, and values forms the foundation of Metrology 5.0, understood and perceived as the metrology of the future. The key factor for success is human-machine collaboration, not human replacement, since it is the human being who will remain at the center of the measurement process. Metrology 5.0 does not eliminate people - it enhances their competencies through tools supported by technologies such as artificial intelligence. It will also strengthen ergonomic and safety aspects, as operators will perform fewer repetitive tasks and more strategic decision-making. Equipped with new skills - such as working with digital measurement systems, understanding statistical models, and applying AI-based procedures - the metrologist of the future will operate in next-generation metrology laboratories that will become centers of knowledge and data analysis, rather than merely facilities for performing measurements.

It is also worth remembering that metrology is a classical example of an interdisciplinary science, combining theory and practice, science and industry, technology and law. Its mission is not only to improve measurement methods but also to build trust in data, upon which modern technological civilization is founded. In this sense, metrology is an indispensable component of scientific progress, innovation, and the information society, and its interdisciplinary nature remains a continuous source of strength and advancement.

References

1. Eichstadt, S., Keidel, A., Tesch, J.: Metrology for the digital age. Meas.: Sens. **18**, 100232 (2021). https://doi.org/10.1016/j.measen.2021.100232
2. Adel, T., Bilson, S., Levene, M., Thompson, A.: Trustworthy Artificial Intelligence in the Context of Metrology. arXiv:2406, 10117 (2024). https://doi.org/10.48550/arXiv.2406.10117
3. Barbosa, C.R.H., Sousa, M.C., Almeida, M.F.L., Calili, R.F.: Smart manufacturing and digitalization of metrology: a systematic literature review and a research agenda. Sensors. **22**, 6114 (2022). https://doi.org/10.3390/s22166114
4. Thompson, A.: Analytical results for uncertainty propagation through trained machine learning regression models. Measurement. **234**, 114841 (2024). https://doi.org/10.1016/j.measurement.2024.114841
5. Samadi, H., Ahsan, M.M., Raman, S.: Metrology and manufacturing-integrated digital twin (MM-DT) for advanced manufacturing: insights from coordinate measuring machine (CMM) and FARO arm measurements. Next Res. **2**(2), 100299 (2025). https://doi.org/10.1016/j.nexres.2025.100299
6. Andolina, M.W.: Youth, generations, and generational research. Polit. Sci. Quart. **139**(2), 281–293 (2024). https://doi.org/10.1093/psquar/qqad079
7. Crescenzo, N.: Young sociologists in the mirror: digital ethnographies of young people online. Front. Sociol. **8**, 1197560 (2023). https://doi.org/10.3389/fsoc.2023.1197560
8. Tirocchi, S.: Generation Z, values, and media: from influencers to BeReal, between visibility and authenticity. Front. Sociol. **8**, 1304093 (2024). https://doi.org/10.3389/fsoc.2023.1304093
9. Koch, T., Laaber, F., Arenas, A., Florack, A.: Socially (dis)connected in a connected world: the role of young people's digital maturity. Comput. Hum. Behav. **163**, 108473 (2025). https://doi.org/10.1016/j.chb.2024.108473
10. Hickman, C., et al.: Climate anxiety in children and young people and their beliefs about government responses to climate change: a global survey. Lancet Planet. Heath. **5**(12), E863–E873 (2021). https://doi.org/10.1016/S2542-5196(21)00278-3
11. Kaman, A., et al.: Youth mental health in times of global crises: evidence from the German longitudinal COVID-19 and psychological health study. J. Adolesc. Health. **1–8** (2025). https://doi.org/10.1016/j.jadohealth.2025.05.010
12. Panchal, U., et al.: The impact of COVID-19 lockdown on child and adolescent mental health: systematic review. Eur. Child Adolesc. Psychiatry. **32**, 1151–1177 (2023). https://doi.org/10.1007/s00787-021-01856-w

13. Thorisdottir, I.E., et al.: Effect of the COVID-19 pandemic on adolescent mental health and substance use up to march, 2022, in Iceland: a repeated, cross-sectional, population-based study. Lancet Child Adolescent Health. **7**(5), 347–357 (2023). https://doi.org/10.1016/S2352-4642(23)00022-6
14. Eichstadt, S., Vedurmudi, A.P., Gruber, M., Hutzschenreuter, D.: Fundamental aspects in sensor network metrology. Acta IMEKO. **12**(1), 1–6 (2023). https://doi.org/10.21014/actaimeko.v12i1.1417
15. Thiel, F., et al.: A digital quality infrastructure for Europe: the European metrology cloud. PTB-Mitteilungen. **4**, 83–97 (2017)
16. Rejeb, A., Rejeb, K., Zrelli, I., Kayikci, Y., Hassoun, A.: The research landscape of industry 5.0: a scientific mapping based on bibliometric and topic modeling techniques. Flex. Serv. Manuf. J. (2024). https://doi.org/10.1007/s10696-024-09584-4
17. Huber, M., Muller, T., Schmitt, R.H.: Differentiation between model error and measurement uncertainty in machine learning modeling for measurement processes. Meas.: Sens. **38**, 101608 (2025). https://doi.org/10.1016/j.measen.2024.101608
18. Kuster, M.: Toward a metrology-information layer for digital systems. Acta IMEKO. **12**(1) (2023). https://doi.org/10.21014/actaimeko.v12i1.1416
19. Oppermann, A., et al.: Digital transformation in metrology: building a metrological service infrastructure. Proc. Comput. Sci. **200**, 308–317 (2022). https://doi.org/10.1016/j.procs.2022.01.229
20. Grasso Toro, F., Lehmann, H.: Brief overview of the future of metrology. Meas.: Sens. **18**, 100306 (2021). https://doi.org/10.1016/j.measen.2021.100306
21. Leach, R., Thompson, A., Senin, N.: A Metrology horror story: the additive surface. ASPEN/ASPE 2017 Spring Topical Meeting on Manufacture and Metrology of Structured and Freeform Surfaces for Functional Applications (2017). https://nottingham-repository.worktribe.com/output/850684
22. Dreyfus, P.A., Psarommatis, F., May, G., Kiritsis, D.: Virtual metrology as an approach for product quality estimation in industry 4.0: a systematic review and integrative conceptual framework. Int. J. Prod. Res. **60**(2), 742–765 (2021). https://doi.org/10.1080/00207543.2021.1976433
23. Wieczorowski, M., Trojanowska, J.: Towards metrology 4.0 in dimensional measurements. J. Mach. Eng. **23**(1), 100–113 (2023). https://doi.org/10.36897/jme/161717
24. Krolczyk, G., Kacalak, W., Wieczorowski, M.: 3D parametric and nonparametric description of surface topography in manufacturing processes. Materials. **14**(8), 1987 (2021). https://doi.org/10.3390/ma14081987
25. Gapinski, B., Wieczorowski, M., Grzelka, M., Alonso, P.A., Tome, A.B.: The application of micro computed tomography to assess quality of parts manufactured by means of rapid prototyping. Polimery. **62**(1), 53–59 (2017). https://doi.org/10.14314/polimery.2017.053
26. Krawczyk, B., Swojak, N., Gapiński, B., Szablewski, P., Wieczorowski, M.: The use of laser triangulation sensor for non-contact measurement of aerospace thread profiles under variable conditions. Measurement. **258**(D), 119410 (2026). https://doi.org/10.1016/j.measurement.2025.119410
27. Maculotti, G., Giorio, L., Genta, G., Galetto, M.: Traceability and uncertainty of defects automated measurements by CNN-powered machine vision systems. CIRP Ann. **74**(1), 661–665 (2025). https://doi.org/10.1016/j.cirp.2025.03.023
28. Wieczorowski, M., Pereira, A., Carou, D., Gapinski, B., Ramírez, I.: Characterization of 5356 aluminum walls produced by wire arc additive manufacturing (WAAM). Materials. **16**, 2570 (2023). https://doi.org/10.3390/ma16072570
29. Rejeb, A., Rejeb, K., Zrelli, I., Sule, E.: Industry 5.0 as seen through its academic literature: an investigation using co-word analysis. Discov. Sustain. **6**, 307 (2025). https://doi.org/10.1007/s43621-025-01166-0

30. Akundi, A., Euresti, D., Luna, S., Ankobiah, W., Lopes, A., Edinbarough, I.: State of industry 5.0 - analysis and identification of current research trends. Appl. Syst. Innov. **5**, 27 (2022). https://doi.org/10.3390/asi5010027
31. Wieczorowski, M., Kucharski, D., Sniatala, P., Pawlus, P., Krolczyk, G., Gapinski, B.: A novel approach to using artificial intelligence in coordinate metrology including nano scale. Measurement. **217**, 113051 (2023). https://doi.org/10.1016/j.measurement.2023.113051
32. Rab, S., et al.: Digital avatar of metrology. Mapan. **38**, 561–568 (2023). https://doi.org/10.1007/s12647-023-00641-1
33. Maculotti, G., et al.: A shared metrological framework for trustworthy virtual experiments and digital twins. Metro. **4**, 337–363 (2024). https://doi.org/10.3390/metrology4030021
34. Senin, N., Leach, R.: Information-rich surface metrology. Proc. CIRP. **75**, 19–26 (2018). https://doi.org/10.1016/j.procir.2018.05.003
35. Poniatowska, M., Mazurkiewicz, D., Sobecki, P.: Digital twin in metrology: opportunities, current implementations and research challenges. Metrol. Hallmark. **6** (2024)
36. Ostrowska, K., Owczarek, D., Gaska, A., Krawczyk, M., Kupiec, R., Sladek, J.: Assumptions of the digital twin in coordinate metrology. Meas.: Sens. **38**, 101667 (2025). https://doi.org/10.1016/j.measen.2024.101667
37. Vlaeyen, M., Haitjema, H., Dewulf, W.: Digital twin of an optical measurement system. Sensors. **21**, 6638 (2021). https://doi.org/10.3390/s21196638
38. Gapinski, B., Wieczorowski, M., Mietlinski, P., Mathia, T.G.: Verification of computed Tomograph for dimensional measurements. In: Diering, M., Wieczorowski, M., Harugade, M., Pereira, A. (eds.) Advances in Manufacturing III. MANUFACTURING 2022. Lecture Notes in Mechanical Engineering, pp. 142–155. Springer, Cham (2022). https://doi.org/10.1007/978-3-031-03925-6_13
39. Berkmans, F., Bigerelle, M., Lemesle, J., Nys, L., Wieczorowski, M., Brown, C.: Peer assessment in interdisciplinary learning: measuring reliability and engaging critical thinking. Think. Skills Creat. **58**, 101950 (2025). https://doi.org/10.1016/j.tsc.2025.101950
40. Wieczorowski, M., Kurkiewicz, A., Krawczyk, J., Krolczyk, J., Szewczyk, P., Gera, B.: A model solution of implementation doctorate as a way to reignite innovation in polish industry. In: Machado, J., Trojanowska, J., Antosz, K., Leão, C.P., Knapcikova, L., Sover, A. (eds.) Innovations in Industrial Engineering IV. Icieng 2025. Lecture Notes in Mechanical Engineering. Springer, Cham (2025). https://doi.org/10.1007/978-3-031-94484-0_18
41. Krawczyk, J., Krolczyk, J., Kurkiewicz, A., Wieczorowski, M.: Development of the "implementation doctorate" program in Poland in the field of engineering and technical sciences on the example of topics related to metallic materials. In: Proceedings of the 33rd International Conference on Metallurgy and Materials, pp. 615–620 (2024). https://doi.org/10.37904/metal.2024.4924

Selection of Factors Affecting Coordinate Measurement Uncertainty in the Context of Designing Database Solutions for Uncertainty Estimation

Danuta Owczarek[1(✉)], Piotr Gąska[2], Michał Jedynak[1], Wiktor Harmatys[1], and Maciej Gruza[1]

[1] Cracow University of Technology, Cracow 31-155, Poland
danuta.owczarek@pk.edu.pl
[2] AGH University of Krakow, Cracow 30-059, Poland

Abstract. The estimation of measurement uncertainty is a critical challenge in metrology. In recent years, research efforts aimed at developing or improving uncertainty estimation methods have increasingly focused on rapidly evolving tools such as artificial intelligence, machine learning, and neural networks, as well as on the development of virtual machines and digital twins. Extensive databases are being developed to serve as the input and foundation for model training. What is crucial here is not only the accurate identification of relevant factors, but also their parametric representation, which enables the determination of their weights with respect to their influence on the output. At the Laboratory of Coordinate Metrology, a project aimed at developing a Machine learning-based expert system that will ultimately constitute a new tool for modelling coordinate measurement uncertainty is currently being carried out. This article presents the selection of factors influencing measurement uncertainty for both optical and tactile coordinate measuring systems, including: the measuring machine, software, object parameters such as size and shape, the measured feature, the measurement method, and the uncertainty estimation method. The methodology and preliminarily evaluated weights of individual factors are also discussed in the context of uncertainty estimation using a statistical analysis – multiple regression. The discussion is preceded by an introduction that reviews classical uncertainty evaluation methods, including the comparative and multi-position methods, as well as virtual and simulation techniques, and the software employed in the uncertainty estimation process. The concept of uncertainty estimation based on archival data and a parametric model is also outlined.

Keywords: Uncertainty Estimation · Artificial Intelligence · Database · Accuracy · Machine Learning · Expert System · Measurement Systems · Parameter Estimation

M. Hryb et al. (Eds.): MANUFACTURING 2026, LNME, pp. 17–31, 2026.
https://doi.org/10.1007/978-3-032-23038-6_2

1 Introduction

1.1 Uncertainty Estimation – Traditional Approaches and Current Challenges

Measurement uncertainty is a key factor in assessing the compliance of products with specifications. It is no longer considered good practice but has become an absolute requirement under ISO/IEC 17025:2017 [1]. In the era of Industry 4.0 and 5.0, both classical and more advanced modern methods of uncertainty estimation are being applied, alongside software tools supporting metrologists. A clear trend can be observed in the development of tools toward digital twins [2, 3] or virtual machines [4, 5].

Classical Methods of Uncertainty Estimation

Among classical methods for uncertainty estimation, a widely used approach is the comparative method defined in ISO 15530-3:2011 [6]. This method involves multiple measurements of a calibrated standard that closely matches the dimensions and shape of the object under examination. It is mainly used in mass inspection of gauges carried out in specialized calibration laboratories and in large-scale production. However, it should be noted that this approach requires an expensive process to ensure high reproducibility of measurement conditions, particularly with regard to temperature control. Despite its widespread use, the comparative method incurs costs related to the need for multiple standards and their calibrations, as well as low versatility and limited applicability for objects with complex geometries.

Another commonly employed technique is the multi-position method (currently under standardization), which is based on measuring the object in various locations and orientations. This method provides a high level of accuracy, but it is time-consuming and requires complex measurement procedures. Only advanced laboratories can afford to perform such studies, as industrial laboratories often lack sufficient standards, financial resources, or personnel to conduct these measurements freely.

Simulation and Virtual Methods for Uncertainty Estimation

Over the past ten years, there has been a noticeable shift toward simulation-based methods, and research has focused on virtual machines that significantly accelerate the measurement accuracy assessment process. A well-established approach to estimating coordinate measurement uncertainty is the application of simulation-based methods. Since the first attempts reported in the early 1990s [7], numerous studies have been published employing different computational strategies and targeting various measurement systems. These methods most frequently rely on performing a series of simulations of the evaluated measurement task using the Monte Carlo Method, enabling uncertainty estimation through subsequent statistical analysis of the simulated results.

A practical implementation of the Monte Carlo approach was presented in [8], where the concept referred to as Expert CMM was described. It incorporates the influence of geometric machine errors on coordinate acquisition for individual measurement points used in the evaluation of defined features. The authors assumed a modular structure of the model to allow future inclusion of additional factors affecting CMM measurement accuracy. A modular architecture also characterizes the uncertainty-estimation model described in [9]. In comparison with earlier models, it additionally considers, among others, probe-related errors and scale errors. Moreover, the method enables the consideration of potential form deviations of the measured part by incorporating a database of measured profiles from representative samples, which allows credible uncertainty estimates even for measurement strategies based on a small number of points.

In [10], the Monte Carlo method is used to simulate CMM errors arising from uncorrected kinematic errors as well as errors connected with probing system orientation. A significant novelty of this model is its adaptation for industrial use by including the influence of temperature on the distribution of residual errors of the simulated device, and thus incorporating these effects into the simulations. An extension of this concept for five-axis systems equipped with continuously indexing probe heads is presented in [11]. The key modification is the development of a module responsible for simulating probing system errors, allowing the model to reproduce variations in error behavior associated with orientation of the probe utilized during measurements.

Another attempt to develop a virtual machine for CMM measurements was described in [12]. Here, simulation required for uncertainty estimation is again carried out using the Monte Carlo method, but the model is simplified such that error simulation relies primarily on calibration-certificate data, especially those describing inaccuracies related to the probing-contact process.

The Monte Carlo method is also used for uncertainty estimation in other coordinate-measuring systems. In [13], a virtual model of an articulated-arm coordinate measuring system is presented, accounting primarily for geometric inaccuracies of the arm structure and encoder errors in rotational joints. By repeatedly simulating the forward-kinematic task while incorporating the modelled error terms, a large set of measurement results can be generated and used to determine uncertainty. Similarly, in the case of laser-tracker systems, the Monte Carlo method can be applied for uncertainty estimation, as demonstrated in [14], where the uncertainty of distance measurements was calculated based on a model of errors associated with interferometric measurements.

Additionally, the literature reports solutions in which the Monte Carlo method is combined with other techniques for modelling measurement uncertainty. In [15], a virtual machine is described in which the error values assigned to simulated repetitions of the measurement sequence are determined using artificial neural networks. Another example can be found in [16] where the Monte Carlo method is combined with error ellipse theory in order to assess the straightness measurements performed with CMM.

The potential application of neural networks in assessing CMM measurement accuracy is also mentioned in [17]. The authors discuss the theoretical basis for using neural networks to generate suitable datasets for Monte Carlo simulations. The behavior of a specific CMM can be represented as a sequence of transformations enabling the computation of measured-point coordinates, while genetic algorithms are proposed as a strategy to increase the efficiency of the training process.

In [18], the authors indicate the possibility of applying machine-learning methods to develop decision-support systems for evaluating measurement processes, building on prior experience and previously acquired measurement results. Also [19] describes how machine learning can be used for metrology digital twin to support error identification and to predict their distribution contributing to better resemblance of actual system by its digital twin. Many examples of usage of AI based solutions used for uncertainty analysis can be found also in [20].

Metrology Software Supporting Uncertainty Estimation

One of the additional challenges in uncertainty estimation and in the broader adoption of simulation-based methods is the difficulty of integrating them with the most widely used metrology software. Commercially available programs such as 3D Reshaper, Focus, or PCDMIS allow the measurement of individual features, often including comparison with a CAD model. Numerous tools of this type exist on the market, sharing similar structures and functionalities, and the segment is largely saturated. However, their purpose is different: they enable the determination of dimensional values, whereas the task of estimating measurement uncertainty remains the responsibility of the operator. Baldwin et al. [21] described the PUNDIT software, which enables the estimation of coordinate-measurement uncertainty. The authors presented the main assumptions and the system architecture based on the simulation by constraints method [22]. In addition to analyzing the applicability of the tool in specific case studies, they identified a set of factors influencing CMM measurement uncertainty, grouping them into effects related to machine geometry, characteristics of the measurement probe, environmental influences, the adopted measurement strategy, and computational algorithms.

In [23], software based on an analytical approach to measurement-uncertainty calculation was presented. The model assumes the determination of uncertainty using a mathematical representation of the measurement task, derived directly from the definitions of tolerances specified in GPS standards. The software allows the inclusion of information about the measurement strategy used for a given inspection task, covering data such as probe-tip selection, the size of controlled features, and the number of points acquired during measurement. Other examples of uncertainty-estimation tools include QMSys GUM Standard, Evaluator version 3.0 (Newton Metrology Ltd), and Metrodata GmbH GUM Workbench Pro [24], whose computational algorithms are predominantly based on the principles outlined in the Guide to the Expression of Uncertainty in Measurement (GUM) [25].

However, it should be noted that effective usage of such softwares often still requires familiarity with complex mathematical formulations and detailed information about machine specific errors, which must be updated after each calibration. In some cases, users are even required to implement additional code segments to complete the analysis. Currently available software only automates certain processes, serving primarily an operational role, while the substantive contribution largely remains with the user. At the Laboratory of Coordinate Metrology, work is underway on a tool that aims to reverse this balance. These efforts are part of the LIDER15/0306/2024 project "An expert system based on machine learning as a new tool for modeling the uncertainty of coordinate measurement" supported by the National Centre for Research and Development (NCBR) under the LIDER program. The outcome of the project will be an application that enables uncertainty estimation in a simple and rapid manner. The developed tool will be practical and useful while remaining compliant with the general requirements of GUM [25], as uncertainties in databases will be calculated based on classical and well-established methods, considering the conformity of the measured task to tasks in the database for which uncertainties have already been estimated. The project assumes that the operator is familiar with the measurement task. Estimation will be possible without performing actual physical measurements.

This article presents the selection of factors influencing uncertainty for both optical and tactile coordinate measurement systems. The information on the factors will provide input to the developing system. Preliminary weights of individual factors in terms of their impact on uncertainty are also indicated. The concept of uncertainty estimation based on archival data is introduced. The novelty of this article lies in the identification of factors in parametric form, along with possible variations, and in indicating the weights of individual components using multiple regression analysis.

2 Concept for Selection and Ranking of Factors Influencing Coordinate Measurement Uncertainty

2.1 Selection and Analysis of Factors Affecting Coordinate Measurement Uncertainty

In model development, the most critical stage is the precise selection of input data, as it forms the basis on which systems learn and generate solutions. Without comprehensive and carefully curated input data, even the most advanced algorithms cannot deliver accurate and reliable results. This issue is particularly pertinent in the context of rapidly evolving artificial intelligence tools, where systems trained on low-quality data tend to produce outputs of limited accuracy.

Research conducted at the Laboratory of Coordinate Metrology (LCM) of Cracow University of Technology involves a wide range of measurements, including the use of standards set (see Fig. 1) to provide input data that are both representative and as diverse as possible.

Fig. 1. Laboratory reference standards set.

In the research, a set of elements with diverse geometries is used: a sphere, an outer cone, an inner cone, a dish, an outer cylinder, an inner cylinder, and a rectangular block.

These objects were made of typical structural materials used in coordinate metrology, such as steel, ceramics, and aluminium.

These databases serve as the foundation for model training. Current measurements are being carried out using multi-sensor systems available in the Laboratory, including the Zeiss O-Inspect Coordinate Measuring System (CMS), the Romer Absolute Arm 7320RI Articulated Arm Coordinate Measuring Machine (AACMM) equipped with both a laser Scanner RS2 and a tactile probe, and the Zeiss WMM850S Coordinate Measuring System (CMS) with a video probe RVP (see Fig. 2). For many years, the Laboratory has been engaged in research aimed at enhancing the accuracy and performance of these systems [26–29].

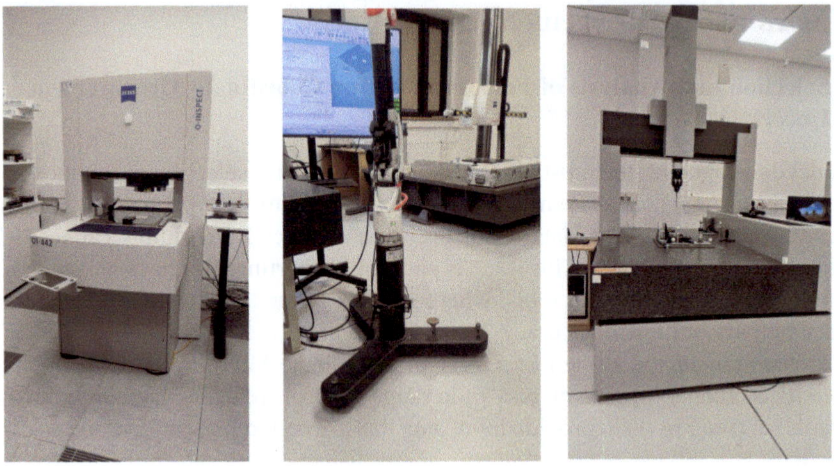

Fig. 2. O-Inspect Zeiss CMS, Romer Absolute Arm 7320RI and Zeiss WMM850S CMS.

2.2 Application of Multiple Regression for Determining the Weights of Selected Factors

The regression tool was selected for the preliminary assessment of the weights of individual factors in the context of the estimated uncertainty of coordinate measurements. The purpose of regression analysis is to examine the relationships between multiple independent (explanatory) variables $\mathbf{x}_1, \mathbf{x}_2, \mathbf{x}_3, \ldots, \mathbf{x}_\mathbf{k}$ and a dependent (response) variable $\mathbf{y}$. Regression allows to identify which $\mathbf{x}$ variables contribute most significantly to explaining the variability of $\mathbf{y}$. It can also be used to observe how the variable $\mathbf{y}$ behaves in response to specific values assumed by the $\mathbf{x}$ variables [30, 31]. These considerations can be expressed in matrix form (1):

$$\begin{bmatrix} y_1 \\ y_2 \\ \vdots \\ y_n \end{bmatrix} = \begin{bmatrix} 1 & x_{21} & x_{31} & \ldots & x_{k1} \\ 1 & x_{22} & x_{32} & \ldots & x_{k2} \\ \vdots & \vdots & \vdots & \vdots & \vdots \\ 1 & x_{2n} & x_{3n} & \ldots & x_{kn} \end{bmatrix} \times \begin{bmatrix} \beta_1 \\ \beta_2 \\ \vdots \\ \beta_k \end{bmatrix} + \begin{bmatrix} \varepsilon_1 \\ \varepsilon_2 \\ \vdots \\ \varepsilon_n \end{bmatrix} \tag{1}$$

or in its abbreviated form (2):

$$\mathbf{y} = \mathbf{X\beta} + \boldsymbol{\varepsilon} \quad (2)$$

$$\mathrm{E}(\mathbf{y}|\mathbf{X}) = \mathbf{X\beta} \quad (3)$$

The expected value of the variable **y** for a given observation matrix **X** is equal to the product $\mathbf{X\beta}$ (3). Populations are often very large, which rarely allows for examining all elements and determining the parameter vector $\boldsymbol{\beta}$ directly. Therefore, it is common practice to draw a random *n*-element sample and to verify whether the relationship holds within it (4), thereby gaining insight into the dependency in the sample [30]:

$$\hat{\mathbf{y}} = \mathbf{Xb} \quad (4)$$

where:

$\hat{\mathbf{y}}$ – column vector, the estimator E(**y**| **X**),
b – column vector (5) determined from the sample, which simultaneously serves as an estimator of β.

$$\mathbf{b} = \begin{bmatrix} b_1 \\ b_2 \\ \vdots \\ b_k \end{bmatrix} \quad (5)$$

The regression coefficients (parameters $\mathbf{b}_1$, $\mathbf{b}_2$, $\mathbf{b}_3$, …, $\mathbf{b}_\mathbf{k}$) represent the weights of the components in the context of measurement uncertainty, assigned to the factors that determine this uncertainty. The presented method also includes the interpretation of the multiple coefficient of determination $\mathbf{R}^2$, the assessment of the statistical significance of individual model variables – if a given model variable is not statistically significant (e.g., according to the Student's t-test), as well as correlation tests [31].

3 Results

The presented results were obtained and developed based on the research and analyses conducted to date (including the multiple regression analysis described above), a review of the literature, and the extensive experience of the operators, encompassing both scientific work and valuable practical measurement experience in industrial settings.

3.1 Factors Affecting the Uncertainty of Coordinate Measurement

In the Laboratory of Coordinate Metrology, the factors influencing the measurement outcome and the associated estimated uncertainty were identified in a general form. These factors can be presented using Ishikawa diagrams (see Fig. 3).

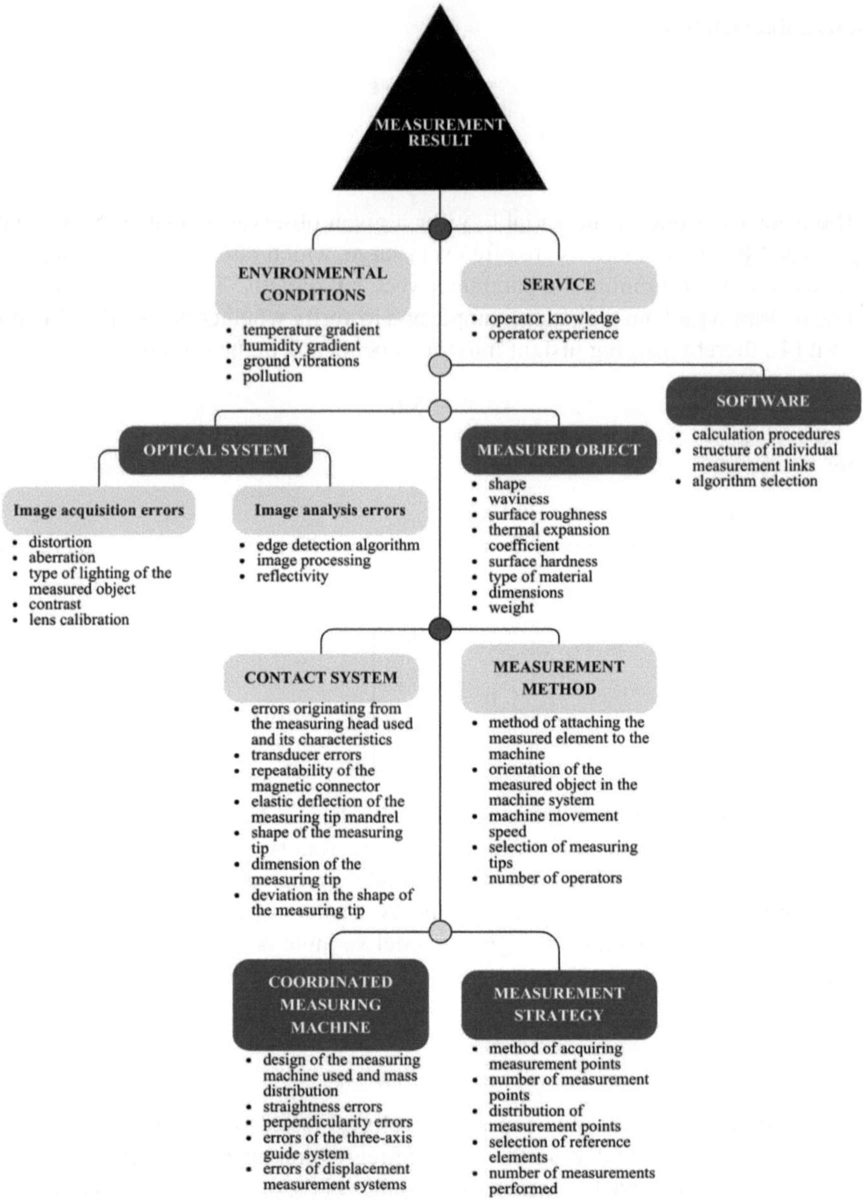

Fig. 3. Ishikawa diagram for factors affecting measurement results.

The factors affecting the uncertainty of coordinate measurements can be expressed in the form of a model [31]:

$$\mathbf{u_i} = \mathbf{f}(\mathbf{x_{1i}}, \ldots, \mathbf{x_{ki}}; \boldsymbol{\beta}_1, \ldots, \boldsymbol{\beta}_\mathbf{k}) + \boldsymbol{\varepsilon}_\mathbf{i} = \tag{6}$$

$$= \boldsymbol{\beta}_1\mathbf{x_{CMMi}} + \boldsymbol{\beta}_2\mathbf{x_{Meti}} + \boldsymbol{\beta}_3\mathbf{x_{FTi}} + \boldsymbol{\beta}_4\mathbf{x_{Geoi}} + \boldsymbol{\beta}_5\mathbf{x_{Di}} + \boldsymbol{\beta}_6\mathbf{x_{Si}} + \boldsymbol{\varepsilon}_\mathbf{i}$$

where:

$\mathbf{x_{CMM}}$ – component from the type of coordinate measuring machine,
$\mathbf{x_{Met}}$ – component from the method of uncertainty estimation,
$\mathbf{x_{FT}}$– component from the measured feature and object type,
$\mathbf{x_{Geo}}$– component from the type of geometric quantity,
$\mathbf{x_D}$ – component from the dimension of the measured object,
$\mathbf{x_S}$ – component from the material of standard,
$\boldsymbol{\beta}_1, \ldots, \boldsymbol{\beta}_k$– coefficients representing the independent contributions of each in dependent variable (component) to the prediction of the dependent variable **u** (uncertainty),
$\boldsymbol{\varepsilon}_i$– random component (model residual) – the intercept term.

Component $\mathbf{x_{CMM}}$ – Type of Coordinate Measuring Machine (CMM). describes the influence of the measurement system employed. The selection of machines in the overall tests includes representative types of CMMs with diverse kinematics, measuring volumes, and probe operation principles, such as: a multisensor CMM, equipped with an optical probe and a white light sensor, a articulated arm with a triangulation laser probe, a laser tracker system. The choice of these classes of machines allows for the reflection of typical industrial scenarios – from precision measurements of small components to measurements of large-scale objects under production conditions. Consequently, this component encompasses both differences in kinematic errors of the machine designs and the metrological properties of optical probes, such as resolution, noise, sensitivity to reflections, and environmental conditions.

In the statistical model, this factor enables the determination of the relative contribution of a specific CMM class to the total variance of measurement uncertainty.

Component $\mathbf{x_{Met}}$ – Method of Uncertainty Estimation. refers to the chosen method of estimating measurement uncertainty. In this study, three approaches were considered: the multi-position method, the comparative method, and simulation-based methods developed at the Laboratory. Each approach places different emphasis on the roles of the machine, the reference standard, and the mathematical model, which consequently leads to varying uncertainty estimates for the same measurement configuration. In the case of the simulation-based method, the critical aspect is the construction of a reliable mathematical model of the measurement system and its errors. Preparing such a model is labor-intensive and requires a thorough understanding of the characteristics of the given machine; however, once defined, the computations themselves are relatively quick. At the same time, even small differences in the metrological parameters of individual CMMs can result in noticeable discrepancies in simulation results if the model is not specifically tailored to the machine in question. The multi-position method does not require an extensive reference standard set but imposes strict requirements on the fixturing and positioning of the object. The comparative method has a different structure: it requires a sufficiently large and well-characterized set of reference standards, while the accuracy of the machine plays a lesser role, primarily serving as a carrier of measurement information between the standard and the tested object. When the conditions for proper standard calibration are met, this method provides very stable and repeatable results. Thus, this component captures the extent to which the choice of uncertainty estimation method affects the resulting combined uncertainty values and may lead to systematic differences between results.

Component x_{FT} – Measured Feature and Object Type. refers to the type of measured feature and the type of measurement object on which this feature is realized. The combination of a specific shape with a given material determines how the feature is represented by the optical system and how the measurement strategy is selected. Elements with high curvature or complex geometry (e.g., the dish, cones, inner cylinders) are sensitive to the measurement path, occlusion of surface fragments, and local variations in lighting conditions. Material properties, in turn, influence the nature of light reflection and edge contrast, which affects the uncertainty of determining geometric parameters. The component therefore groups these effects, assigning them a separate contribution within the uncertainty model, resulting from the combination of the geometry and material of the measured object.

Component x_{Geo} – Type of Geometric Quantity. is associated with the type of geometric quantity of measured object being analyzed. The study considered, among others, length, angle, straightness, flatness, and parameters describing the shape of an element. Each of these quantities requires a different approach both at the data acquisition stage and during subsequent data processing. For length measurement, a limited set of points, e.g., point-to-point along the measurement axis may be sufficient, whereas the assessment of straightness or flatness requires denser and appropriately distributed sampling of the entire surface or segment. The set of applied algorithms also varies: from least-squares fitting of a line or a plane, through form-filtering procedures, to more complex geometric constructions used, for example, in angle determination.

Component x_D – Dimensions of the Measured Object. describes the influence of the object's overall dimensions on the measurement uncertainty. The analysis considered several length intervals corresponding to different geometric scales: <0.1 mm; 1 mm), <1 mm; 10 mm), <10 mm; 50 mm), <50 mm; 100 mm), <100 mm; ∞). As the characteristic dimensions of the object increase, the contribution of kinematic errors of the measurement system grows, as does the significance of environmental effects such as temperature variations, vibrations, or air movement. Additionally, the impact of thermal expansion, both of the measured object and of the components of the measurement system becomes more pronounced. This component enables a quantitative description of the relationship between the size of the object and the level of uncertainty with which its geometric parameters are determined.

Component x_S – Material of Standard. refers to the material from which the reference standard is made. The analysis included steel standards, ceramic standards, and standards coated with a TiN (titanium nitride) layer, which is used to reduce surface reflectivity. The choice of material affects both the interaction of the measurement beam with the surface and the dimensional stability of the standard under typical laboratory conditions. For optical measurements, the reflective properties of the surface are of particular importance, including the level of reflectance, the uniformity of the top layer, and the directionality of reflection. At the same time, the material of the standard, through its coefficient of thermal expansion, determines dimensional changes with ambient temperature and influences the susceptibility of the surface to degradation processes such as oxidation or scratching. The use of standards made of different materials and coatings therefore leads to varying optical measurement conditions, which directly affects the uncertainty of the determined geometric parameters. This component thus allows the statistical model to separate this effect from the general influence of the object's geometry and the characteristics of the measurement system used.

3.2 Weights of Factors Influencing Coordinate Measurement Uncertainty – Preliminary Analysis

To determine the influence strength the weights of individual factors on measurement uncertainty, it was necessary to assemble a sample of results for selected features in the form of a dedicated database. Initially, the dataset contained uncertainty results calculated using three uncertainty estimation methods (the comparative method, the multiposition method, and the method developed in the Laboratory). Detailed data and a description of the measurement systems are provided in [31].

The database was supplemented with additional records containing uncertainty information sourced from the LCM archive and from complementary measurements. The database was constructed in such a way that individual records reflected the combinations of factor variants most commonly encountered in practice. In principle, the most reliable regression results are obtained when the dataset is as large and as diverse as possible. Therefore, nearly all available information that had been verified for correctness and was suitable for use as input data was included. The data were prepared in a form enabling regression analysis using 0/1 encoding (see Fig. 4).

Uncertainty	Method	Characteristic	Object	Size	Standard	Machine	Comparative	Multiposition	OPTI-U	Length	Shape	Sphere	Cone	From 1
0.0239	Comparative	Length	Sphere	25	Sphere	CMM	1	0	0	1	0	1	0	0
0.0255	Comparative	Length	Sphere	25	Sphere	CMM	1	0	0	1	0	1	0	0
0.0143	Multiposition	Length	Sphere	25	Sphere	CMM	0	1	0	1	0	1	0	0
0.0102	Multiposition	Length	Sphere	25	Sphere	CMM	0	1	0	1	0	1	0	0
0.0323	Multiposition	Length	Sphere	25	Sphere	CMM	0	1	0	1	0	1	0	0
0.0209	OPTI-U	Length	Sphere	25	Sphere	CMM	0	0	1	1	0	1	0	0
0.0110	OPTI-U	Length	Sphere	25	Sphere	CMM	0	0	1	1	0	1	0	0
0.0096	OPTI-U	Length	Sphere	25	Sphere	CMM	0	0	1	1	0	1	0	0
0.0257	Comparative	Length	Sphere	25	Sphere	CMM	1	0	0	1	0	1	0	0
0.0329	Comparative	Length	Sphere	25	Sphere	CMM	1	0	0	1	0	1	0	0
0.0146	Multiposition	Length	Sphere	25	Sphere	CMM	0	1	0	1	0	1	0	0
0.0094	Multiposition	Length	Sphere	25	Sphere	CMM	0	1	0	1	0	1	0	0
0.0228	Multiposition	Length	Sphere	25	Sphere	CMM	0	1	0	1	0	1	0	0

Fig. 4. Conversion to 0/1 format.

The considerations so far have allowed the formulation of information on the influence of the given factors in the form of Table 1.

Table 1. Preliminary estimation of weights for factors affecting measurement uncertainty

Factor	Coefficient value	R2	Coefficient × R2	Lower 95%	Upper 95%	Impact on Uncert. (%)
U – method	0.038	91	3.467	0.01547041	0.05823326	6
Feature	0.107	91	9.764	0.07755723	0.22341671	18
Object	0.295	91	26.818	−0.36254820	−0.15758779	49

(continued)

Table 1. (*continued*)

Factor	Coefficient value	R2	Coefficient × R2	Lower 95%	Upper 95%	Impact on Uncert. (%)
Object size	0.040	91	3.595	−0.00018610	−0.00000003	7
Standard	0.014	91	1.256	0.00000189	0.02680371	2
Machine	0.053	91	4.823	−0.06503663	−0.03424647	9

The database consists of input data x, which include the components presented in Table 1 and their corresponding variants, e.g., measurement method (comparative, multiposition, Opti-U), as well as output data y – measurement uncertainty. The data were presented using 0/1 encoding, except for the object size, for which the numeric value was retained. The model was developed under the assumption that the number of observations must be greater than the number of model parameters. Therefore, the database contained 58 different combinations. Statistical significance was assessed using the Student's t-test – the approach and hypotheses are described in detail in [31]. The significance levels obtained are presented in Table 1. The significance analysis allowed for the removal of two variants, as they showed negligible statistical importance. Additionally, correlation tests were conducted, which revealed a clear positive relationship (value "1") between certain variants, e.g., "flatness standard" and "flatness." This indicates that the occurrence of the "flatness standard" variant guarantees 100% the occurrence of the "flatness" variant, allowing for the reduction of columns to one without losing information. The SS parameter (Sum of Squares), which measures the total variability in the data, was 0.073648 for regression and 0.007733 for residual variability. The model was validated by adding successive observations – similar values were obtained, indicating the model's stability.

4 Discussion

In this article, emphasis is placed on providing as comprehensive characterization as possible of the factors influencing measurement uncertainty. These factors have been defined in a comprehensive manner nevertheless, only further analyses and model training will allow determining whether any factor that could be considered significant in the context of the study has been overlooked. All solutions developed in the Laboratory will undergo extensive verification and validation [32].

The established weights constitute an initial step in the analysis and confirm that the identified factors indeed influence measurement uncertainty. Nevertheless, the authors are aware that the regression tool selected is a basic one. In further analyses, the factors will be investigated using more advanced methods. Regarding the specific values of the obtained weights (Table 1), the authors emphasize that they were determined for a sample of results from a particular laboratory. Under different conditions, these proportions will shift, and the corresponding effects will have different weights. In the presented example, it was determined that the weights account for 91% of the variability of the dependent variable **y** – measurement uncertainty; however, these data should be regarded as an example rather than a universal proportion. The authors expect a greater contribution from the factor related to the measurement machine.

The authors aim to develop a tool for uncertainty estimation that is as universal as possible, regardless of the system used. Nonetheless, they recognize that, given the specific characteristics of a particular measurement machine, predicting uncertainty based on historical data may be more or less challenging. The team will strive to address the indicated uncertainties as comprehensively as possible.

5 Conclusions

Knowledge of the factors influencing uncertainty and awareness of the significance of their effects is crucial for the efficient planning and optimization of measurement processes. Understanding what is critical in a given situation allows for informed decisions regarding the allocation of tasks and resources to control a specific factor. It may turn out that minor modifications to the measurement task such as the choice of a reference standard or measurement strategy can be more effective than investing in advanced equipment.

In the context of uncertainty estimation, it has been shown that there is still a lack of solutions on the market that allow for rapid uncertainty estimation. The time required to estimate uncertainty with currently available methods is long, and the process itself is considered complex. This complexity often leads to practices in which uncertainty values are assigned in a very simplified manner (for example, quickly adopting the MPE – Maximum Permissible Error as the uncertainty value). Such practices can result in errors when assessing compliance with specifications, potentially causing serious mistakes, claims, and, in extreme cases, catastrophic failures.

The tool developed at the Laboratory of Coordinate Metrology for measurement uncertainty estimation aims to address these challenges and provide the user with a practical, easy-to-use solution, transforming uncertainty estimation in industrial practice from a complex and rarely addressed issue into a standard procedure.

Acknowledgements. This study was supported by the National Center for Research and Development, Poland, under the LIDER program (grant no. LIDER15/0306/2024). The authors also acknowledge the Laboratory of Coordinate Metrology for providing access to the measurement facilities.

References

1. ISO/IEC 17025: General requirements for the competence of testing and calibration laboratories (2017)
2. Poroskun, I., Rothleitner, C., Heißelmann, D.: Structure of digital metrological twins as software for uncertainty estimation. J. Sens. Sens. Syst. **11**, 75–82 (2022)
3. Wright, L., Davidson, S.: Digital twins for metrology; metrology for digital twins. Meas. Sci. Technol. **35**(5), 051001 (2024)
4. Ostrowska, K., Gąska, A., Kupiec, R., Gromczak, K., Wojakowski, P., Sładek, J.: Comparison of accuracy of virtual articulated arm coordinate measuring machine based on different metrological models. Measurement. **133**, 262–270 (2019)

5. Wieczorowski, M., Trojanowska, J.: Towards metrology 4.0 in dimensional measurements. J. Mach. Eng. **23**(1), 100–113 (2023)
6. ISO 15530-3: Geometrical product specifications (GPS) — Coordinate measuring machines (CMM): Technique for determining the uncertainty of measurement — Part 3: Use of calibrated workpieces or measurement standards (2011)
7. Kunzmann, H., Trapet, E., Wäldele, F.: Concept for the traceability of measurements with coordinate measuring machines. In: Ikawa, N., Shimada, S., Moriwaki, T., McKeown, P.A., Spragg, R.C. (eds.) International Progress in Precision Engineering, pp. 40–52 (1993)
8. Balsamo, A., Di Ciommo, M., Mugno, R., Rebaglia, B.I., Ricci, E., Grella, R.: Evaluation of CMM uncertainty through Monte Carlo simulations. CIRP Ann. **48**(1), 425–428 (1999)
9. Kruth, J.P., Van Gestel, N., Bleys, P., Welkenhuyzen, F.: Uncertainty determination for CMMs by Monte Carlo simulation integrating feature form deviations. CIRP Ann. **58**(1), 463–466 (2009)
10. Gąska, A. et al.: Virtual CMM model adapted for usage in industrial conditions. In: Proceedings of the IMEKO XXI World Congress Measurement in Research and Industry Proceedings, pp. 1425–1429. IMEKO International Measurement Confederation, Prague Czech Republic (2015)
11. Gąska, P., Gąska, A., Sładek, J., Jędrzejewski, J.: Simulation model for uncertainty estimation of measurements performed on five-axis measuring systems. Int. J. Adv. Manuf. Technol. **104**, 4685–4696 (2019)
12. Li, H.L. et al.: Uncertainty modeling and evaluation of CMM task oriented measurement based on SVCMM. Meas. Sci. Rev. **17**, 226–231 (2017)
13. Ostrowska, K., Gąska, A., Sładek, J.: Determining the uncertainty of measurement with the use of a virtual coordinate measuring arm. Int. J. Adv. Manuf. Technol. **71**(1–4), 529–537 (2014)
14. Gruza, M., Gąska, P., Harmatys, W., Gąska, A.: New method for testing the laser tracker length measuring unit using a large-scale horizontal arm CMM and its application for assessing the accuracy of distance measurements. Precis. Eng. **94**, 54–66 (2025)
15. Sładek, J.: Coordinate Metrology – Accuracy of Systems and Measurements. Springer, Berlin, Heidelberg (2016)
16. Zhu, M., Ge, G., Yang, Y., Du, Z., Yang, J.: Uncertainty evaluation of straightness in coordinate measuring machines based on error ellipse theory integrated with Monte Carlo method. Meas. Sci. Technol. **31**, 035008 (2020)
17. Raya, F.J., Cuesta, E., Martínez-Pellitero, S., Barreiro, J.: Uncertainties on CMMs by applying a model of corrections based on a global metric, Monte Carlo and neural network methods. Procedia Eng. **132**, 796–803 (2015)
18. Olsson, E., Funk, P., Andersson, A.: Case-based reasoning applied to geometric measurements for decision support in manufacturing. Int. J. Syst. Assur. Eng. Manag. **4**, 223–230 (2013)
19. Chen, G., Mehdi-Souzani, C., Anwer, N.: Error source identification in metrology digital twin systems using machine learning. Adv. Eng. Inform. **70**, 104142 (2026)
20. Xia, H. et al.: Measurement uncertainty evaluation with small samples: A review and prospect. Measurement. **258**, 119031 (2026)
21. Baldwin, J.M., Summerhays, K.D., Campbell, D.A., Henke, R.P.: Application of simulation software to coordinate measurement uncertainty evaluations. NCSLI Measure. **2**(4), 40–52 (2007)
22. Phillips, S.D. et al.: The calculation of CMM measurement uncertainty via the method of simulation by constraints. In: Proceedings of the 12th Annual Meeting of the American Society for Precision Engineering, pp. 443–446. American Society for Precision Engineering, Norfolk VA United States (1997)
23. Jakubiec, W., Płowucha, W.: First coordinate measurements uncertainty evaluation software fully consistent with the GPS philosophy. Procedia CIRP. **10**, 317–322 (2013)

24. Jurado, J.M., Alcázar, A.: A software package comparison for uncertainty measurement estimation according to GUM. Accred. Qual. Assur. **10**, 373–381 (2005)
25. Joint Committee for Guides in Metrology: Evaluation of measurement data — guide to the expression of uncertainty in measurement. JCGM. **100**, 2008 (2008) (GUM 1995 with minor corrections)
26. Gąska, A. et al.: Analysis of changes in coordinate measuring machines accuracy made by different nodes density in geometrical errors correction matrix. Measurement. **68**, 155–163 (2015)
27. Sładek, J., Ostrowska, K., Gacek, K.: Kinematic metrological model of the coordinate measuring ARM (MCMA). In: 19th Imeko World Congress 2009, pp. 2618–2623 (2009)
28. Tomczyk, K., Ostrowska, K.: Procedure for the extended calibration of temperature sensors. Measurement. **196**, 111239 (2022)
29. Ostrowska, K., Gąska, A., Kupiec, R., Sładek, J.: Accuracy measuring arms using lasertracer system. In: 11th Imeko TC14 International Symposium on Measurement and Quality Control ISMQC, pp. 98–101 (2013)
30. Górecki, B.R.: Podstawowy kurs nowoczesnej ekonometrii. Key Text (2015)
31. Owczarek, D.: Uncertainty Model in Optical Coordinate Measurement. Monograph Cracow University of Technology, Cracow (2017)
32. Gromczak, K. et al.: Validation model for coordinate measuring methods based on the concept of statistical consistency control. Precis. Eng. **45**, 414–422 (2016)

Advanced Techniques and Methodologies for Surface and Geometry Measurement

Methodology for Assessing the Repeatability of Focal Positioning in Focus Variation Microscopy (FV)

Karol Grochalski(✉), Julia Radziejewska, and Krzysztof Kupka

Faculty of Mechanical Engineering, Poznan University of Technology, Poznan, Poland
karol.grochalski@put.poznan.pl

Abstract. The paper presents a methodology for evaluating the repeatability of objective positioning in the Focus Variation (FV) technique, using advanced procedures for topographic data extraction and their functional analysis. A measurement scheme is proposed based on statistical analysis of two acquisition types: a reference measurement, in which the objective is not displaced by more than 4 mm, and a sequential measurement, in which each subsequent scan is preceded by a displacement of the objective along a selected axis in the range of 40–80 mm. For both strategies, the scatter of surface topography parameters was determined and used to estimate the focal positioning error of the objective, treated as a component of the FV technique's uncertainty budget. It was demonstrated that, in the reference configuration, a repeatability of 98.7% was achieved for key functional parameters, whereas in the sequential mode the repeatability was 77.6%, indicating a significant influence of long axial displacements on positioning stability. The practical implications of these differences for the selection of scanning strategies, comparability of results, and the design of control procedures in surface engineering are discussed. The proposed methodology constitutes a universal tool for the validation of FV systems and can be directly used to monitor the condition of the drive system, optics, and reconstruction algorithms, as well as to quantify measurement uncertainty in industrial and research applications.

Keywords: Focus Variation · Surface Topography · Measurement Repeatability

1 Introduction

The geometrical structure of a surface is one of the key aspects of the manufacturing process, as it is directly related to the efficiency and operational performance of mating components. A particularly important indicator in its assessment is the surface roughness parameter, which describes the quality of the machined surface [1, 2]. Surface topography is not merely a qualitative descriptor, but a feature that determines the functional properties of a component [3]. Pawlus et al. indicated that the Surface Geometrical Structure (SGP) influences tribological phenomena such as friction, wear, and lubrication conditions [3]. Fetecau and Stan likewise confirmed that surface roughness is a quality characteristic which, in most cases, constitutes a technical requirement,

M. Hryb et al. (Eds.): MANUFACTURING 2026, LNME, pp. 35–48, 2026.
https://doi.org/10.1007/978-3-032-23038-6_3

affecting friction and heat transfer [3]. With technological progress, industry has been shifting from traditional two–dimensional (2D) profile analysis towards three–dimensional (3D) characterization [4]. The importance of areal parameters was emphasized by Brown et al., who highlighted the advantages of multi–scale analysis for identifying strong correlations between surface topography and its functional performance [4]. Two main approaches to surface measurement can be distinguished: contact and optical methods [5]. Although contact methods are well standardized, they are time–consuming and carry the risk of damaging the surface due to the interaction with the stylus tip [6, 7]. In contrast, the applicability of optical methods is determined by specific measurement conditions. They offer high measurement speed and do not interfere with the examined surface structure [8]. On the other hand, they are more sensitive to illumination conditions and often exhibit lower repeatability than contact instruments [8, 9]. The reliability of measurement systems therefore becomes a critical issue. Two fundamental indicators used in this context–repeatability and reproducibility–make it possible to assess the quality of the acquired data [6, 10, 11]. Wieczorowski et al. pointed out that, in industrial practice, the results of roughness measurements of the same component often differ between supplier and customer [12]. This discrepancy is related to reproducibility, which Zanobini et al. define as the variability arising from measurements performed by different operators or in different laboratories [11]. The present work, however, focuses on repeatability, understood as the variability observed when a single operator carries out multiple measurements of the same part. Performing a Gage Repeatability and Reproducibility (Gage R&R) analysis makes it possible to assess the capability of a given measurement system [13]. As noted by Razumić et al., it provides indicators of the ability to achieve consistent results and constitutes an essential component of measurement uncertainty, a methodology that is standardized, among others, in ISO 5725 [14].

The accuracy and repeatability of surface topography measurements are influenced by the calibration procedures of the instruments, their stability over time, the analysis and evaluation of measurement data, as well as external factors. In addition, the preparation of measurement data–their quality and the adopted measurement settings–is of key importance [15–17]. To achieve the highest production quality, it is necessary to control and calibrate measuring instruments, for which international standards concerning the measurement and evaluation of surface irregularities and good–practice guidelines are extremely useful [18]. What is required is not only the integration of error–compensation models, but also rigorous calibration in accordance with international standards and norms [19, 20]. Reliable topography measurements can be obtained, among other means, through proper calibration of the height axis of optical instruments. Eifler et al. presented an approach to improving Z–axis calibration of optical measuring devices, in which calibration artefacts are based on the Abbott curve, enabling the visualization of desired surface properties with less effort compared to other methods [21]. Haitjema and co–workers also proposed interferometric methods [22]. Vrba et al. indicated that the scanning strategy, in particular the scanning speed and sampling density, affects measurement uncertainty [23]. Jamuła et al. arrived at similar conclusions, demonstrating that sampling density (k) and the number of measurement tracks (w) have a very strong effect on the repeatability of non–contact measurements [24]. Peta et al. proposed a method for verifying the repeatability and reproducibility of measurements by comparing height values (z) measured at the same positions (x, y), which are then plotted on Height–Height (H–H) plots. These plots provide a graphical representation of the relationship between topographical height measurements obtained without changing the location of the surface with respect to the measuring instrument [25]. New analytical approaches make it possible to detect errors that are not visible in standard Gage R&R analysis. Using multi–scale analysis, Gogolewski observed that repeatability errors vary with scale and that small shifts of morphological features occur between subsequent measurements [10].

The present study focuses on the Alicona μCMM system, which operates on the Focus Variation (FV) technique. Danzl et al. described this technique as a promising solution for measuring roughness, form, and wear [26]. The authors noted that FV exhibits effects similar to those of traditional contact measurements. The use of this method enables the measurement of steep surface slopes, which is difficult with other 3D measurement techniques [27]. By contrast, Mital and Spišák highlighted difficulties in measuring shiny, highly reflective surfaces [5]. Grochalski et al. investigated the influence of illumination direction on the obtained parametric values. Their study demonstrated that both the type and direction of illumination affect the values of surface parameters. Moreover, they showed that the orientation of a sample with an anisotropic structure relative to the polarization axis of the light significantly influences the accuracy of the resulting surface measurements [27]. Harmatys et al. also demonstrated that illumination intensity in optical systems affects measurement accuracy [28]. Interesting conclusions were reported by a team led by Niesłony [35], who indicated that the wavelength of light used in non–contact surface measurements directly influences the results and the graphical representation of surface irregularities. In a subsequent study on non–contact surface measurements, Grochalski et al. showed that proper preparation of measurement data has a decisive impact on the final values of the evaluated parameters [16]. The quality of the acquired measurement data is directly dependent on the measurement system used, in particular on the type and direction of illumination, as well as on temperature [13]. Temperature is identified as one of the key factors influencing the accuracy of broadly understood coordinate metrology (CMM) [29]. Guillory et al. reported that temperature gradients also affect multilateration systems [30]. Ambient temperature and thermal gradients are sources of error that cause deformation of the measuring device due to thermal expansion [31]. As shown by Šafarič et al., they constitute a major source of measurement uncertainty [32]. Optical instruments additionally possess internal heat sources such as illumination, power supplies, and motors. Harmatys et al. and Grochalski et al. presented data indicating that machine warm–up time is critical and introduces a noticeable change in measurement uncertainty [33, 34]. Furthermore, Grochalski et al. demonstrated that the temperature stabilization time should be determined individually, depending on the measuring device. In their studies, the authors proved that a properly thermally stabilized machine can reduce measurement errors by up to 90%.

Despite the extensive research on environmental and optical factors influencing measurement uncertainty, a specific gap remains regarding the repeatability of the objective's mechanical positioning. While standard calibration procedures and classical Gage Repeatability and Reproducibility (Gage R&R) analyses effectively assess the overall capability of a measurement system, they typically aggregate various error sources without isolating the specific contribution of long-range axial displacements [38–40]. Consequently, these established approaches may overlook positioning instabilities that occur specifically during sequential measurements. Furthermore, although modern multi-scale analyses allow for the detection of subtle morphological shifts, they do not explicitly correlate these deviations with the functional stability of the focal drive system. Therefore, a methodology is needed to quantify the impact of positioning errors on functional surface parameters, distinct from optical artifacts.

1.1 Objective and Motivation

To address the identified limitations regarding mechanical positioning stability, the aim of the present study was to determine the repeatability of measurements performed using Focus Variation (FV) microscopy on the basis of a parametric representation of the geometrical surface structure, and to assess the repeatability of measurement beam/objective positioning in a sequential measurement scheme, as well as to identify factors influencing the stability and reliability of the obtained results.

Combining non–contact FV measurements with independent interferometric positioning measurements allows the separation of uncertainty sources–those arising from the topography acquisition process itself from those associated with the mechanics and control of the objective position. This approach enables a reliable evaluation of measurement repeatability, both in terms of the parametric representation of the geometrical surface structure and the stability of beam/objective positioning within the measurement sequence. The results of the analysis are used to formulate recommendations regarding measurement strategies for multi–repeat surface texture investigations.

2 Methodology

2.1 Materials and Methods

Three surfaces made of 7075 aluminium alloy were subjected to examination. The specimens were manufactured so as to ensure the highest possible parallelism and flatness of the surfaces. In addition, prior to the final texturing process, the surfaces were freed from any artefacts originating from previous machining operations. The target texturing was performed by glass bead blasting in order to obtain an anisotropic surface structure. Each surface was characterized by distinct spatial parameters, controlled by the working pressure of the blasting medium: 2 bar, 4 bar, and 6 bar. As the medium, irregular grains with an average particle size of 0.3 mm were used.

2.2 Instrumentation and Data Processing

The topography of the examined surfaces was recorded using an Alicona μCMM measuring system operating with the Focus Variation (FV) technique. This method enables non–contact acquisition of three–dimensional topographic data, allowing the simultaneous reconstruction of complex height and spatial features of the surface at the micro–scale. In contrast to stylus profilometry, the FV technique eliminates the influence of probe–induced deformation and filtering effects at the micro– and nano–scales, so that the acquired data more faithfully represent the true surface morphology. For the purposes of parametric analysis, a representative measurement area of 3.8 mm × 3.8 mm was selected. Data acquisition was performed at 20× magnification; the vertical (Z) resolution was 70 nm and the lateral resolution was 1 μm. This configuration ensures the detectability of the expected surface irregularities and enables both quantitative and qualitative characterization of features such as peaks, valleys, and the spatial directionality of the structure.

The specimens were mounted on the microscope stage parallel to the X–axis of the instrument, at a fixed distance of 40 mm with respect to the measured surfaces. The measurement setup, together with the arrangement of the examined objects, is shown schematically in Fig. 1. This configuration was adopted to verify the repeatability of objective/beam positioning (for the selected X–axis) for successive areas in the measurement sequence; the principle of the positioning sequence is summarized in Table 1.

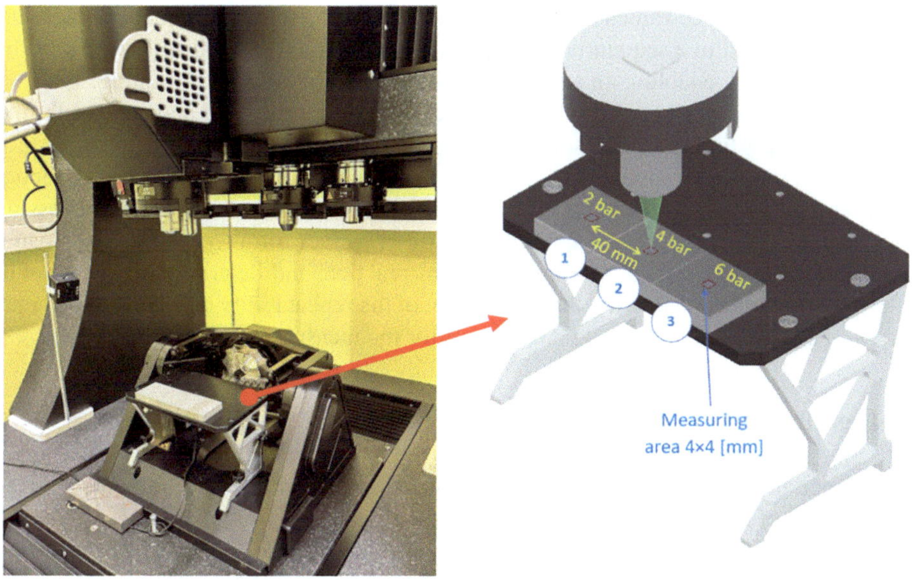

Fig. 1. Arrangement of the measurement surfaces on the measuring stage of the FV microscope.

The figure presents the arrangement of the measurement surfaces on the measurement stage of the Alicona μCMM focus variation microscope. The displacement measurements were carried out with respect to the global coordinate system of the FV microscope. It should be noted that the displacement along the Z–axis is significantly smaller and, in the present study, did not exceed 200 μm. Moreover, potential positioning errors in the Z–axis are likely to have a smaller impact on data acquisition, since the acquisition itself is based on evaluating focus intensity from a series of recorded images [36, 37].

Table 1. Sequence of objective positioning during repeated measurements.

Replicates	Position		
30 (reference)	1	–	–
30 (reference)	–	2	–
30 (reference)	–	–	3
30	1	2	3
30	2	3	1
30	3	1	2

The data presented in Table 1 represent the sequence and number of measurement repetitions for the anisotropic surfaces. To ensure the reliability of the analysis, a statistical evaluation was carried out. First, each selected surface was measured 30 times without rapid or pronounced displacement of the objective (Ref). These measurements constitute the reference values for the selected areal parameters. Subsequently, each surface was measured in a sequential manner (Seq), so that a distinct displacement of the objective occurred, enabling an analysis of the repeatability of its positioning.

After acquisition, all recorded topographic fields were processed and analysed using MountainsMap® Premium 10.3 software. This platform enabled the extraction of surface parameters (amplitude, spatial, functional parameters, etc.), visualization of the topography, and preparation of data for further correlation analyses. At the data–processing stage, standard procedures were applied: removal of surface form errors, leveling, spatial filtering in accordance with good metrological practice, and determination of a set of descriptive parameters allowing the assessment of the repeatability of FV measurements and the repeatability of objective positioning for the measurement locations.

3 Results and Discussion

A key stage in the repeatability analysis of surface topography measurements was the development of a procedure enabling the extraction of representative surface features. For both the reference and sequential measurement modes, an algorithm was prepared to return data that directly capture the relationship between objective (focus) positioning and areal parameters.

Four key parameters were selected for the analysis:

- Sa – arithmetical mean height of the surface,
- Sq – root mean square height of the surface,
- Ssk – skewness of the height distribution,
- NM – fraction (e.g. in %) of topography points for which the system did not record reliable data (no signal, excessive slope, specular reflections), serving as an indicator of data quality and completeness. This parameter should be treated as a very important indicator of repeatability in a qualitative sense. In principle, if the repeatability of focal positioning and the measurement data were free from all possible errors,

this parameter should be NM = 0. The higher its value, the more measured points have no common part (i.e., cannot be matched/overlapped), which in turn indicates a deterioration in the repeatability of the measurement discussed.

The parameter Ssk is particularly important, as its variation reflects changes in the character of the surface structure. The following criteria were adopted:

- Ssk ≈ 0: the height distribution is approximately symmetric (Gaussian). The surface exhibits a comparable number of peaks and valleys, with no pronounced dominance of either.
- Ssk < 0 (negative): the distribution indicates the presence of deep valleys, which may be interpreted as a relatively smooth plateau with deeper pockets/depressions.
- Ssk > 0 (positive): the distribution indicates a larger number of high peaks, and the surface is dominated by protruding asperities.

The above criteria affect the applicability of this parameter for identifying the overall character of the height distribution of peaks and valleys. In the analysis performed, this made it possible to estimate to what extent the positioning of the objective influenced the smoothing of peaks and the flattening of valleys.

The data–extraction algorithm consisted of three main operations:

- determination of parameters for any measured surface from the population (Ref_0),
- computation of mean parameter values (M) from all measured surfaces (also used as an averaged graphical representation of the structure),
- calculation of the absolute difference between surfaces combined with an analysis of non–measured points (NM); the surfaces subjected to mutual subtraction were selected based on the maximum value of the Sq parameter (lowest correlation) among all possible permutations within the population.

A graphical representation of the individual operations is shown in the form of resulting surfaces – a selected representative example for the 4 bar sequential measurement is presented in Fig. 2.

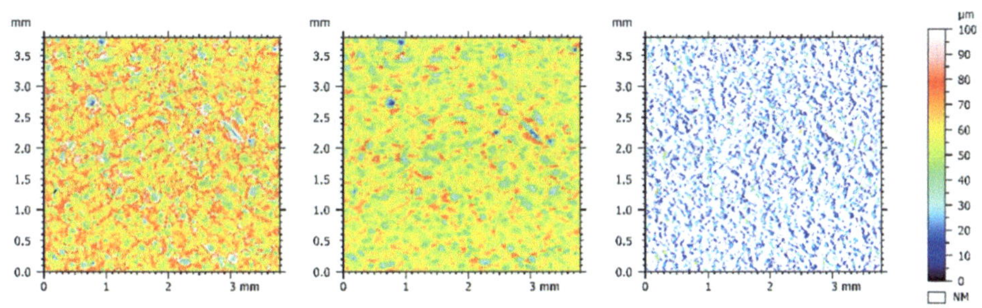

a) b) c)

Fig. 2. Resulting surfaces obtained during functional feature extraction: (a) reference surface, (b) averaged surface (M), (c) visualization of remaining points (differentiating areas).

Figure 2 illustrates the individual operations performed on the data population. Surface (Fig. 2a) represents a single measurement from the population. The determined areal parameters serve as a reference for subsequently evaluated topographic features in the following operations. Surface (Fig. 2b) shows the averaged surface obtained from 30 measurements. This representation is particularly important, as lower similarity indicates lower measurement repeatability. Illustration (Fig. 2c) presents the remaining points after the surface–subtraction operation, for which the Sq parameter was the highest among all possible permutations.

The data identifying a single reference surface (Ref_0) and the averaged surface (M) for the reference (Ref) and sequential (Seq) measurements are summarized in Table 2.

Table 2. Summary of parameters identifying the surface.

	2 bar			4 bar			6 bar		
	Ref_0	M_{Ref}	M_{Seq}	Ref_0	M_{Ref}	M_{seq}	Ref_0	M_{Ref}	M_{Seq}
Sq [μm]	7.79	7.89	6.04	10.50	10.60	8.57	12.40	12.40	11.50
Ssk [−]	−0.09	0.12	−0.32	−0.24	−0.24	−0.37	−0.19	−0.19	−0.26
Sa [μm]	5.99	6.01	4.59	8.20	8.22	6.53	9.65	9.67	8.88
NM [%]	3.05	2.02	0.02	11.10	9.07	0.02	9.53	7.92	0.13

The table presents the values of parameters characterizing the averaged topography of the examined surfaces (Sq, Sa) for the considered measurement strategies, as well as the indicator parameters Ssk and NM. When analysing Sq and Sa with respect to the reference surface for the mean data in the reference measurements (with the objective displacement kept as small as possible), only minor differences were observed, not exceeding $\Delta Sq = 1.3\%$ and $\Delta Sa = 0.4\%$. This indicates a high repeatability of the microscope objective focal positioning. In the case of sequential measurements (objective motion in the range of 40–80 mm), the differences between these parameters increase significantly and amount to $\Delta Sq = 22.4\%$ and $\Delta Sa = 23.4\%$, respectively. The reduction in Sq and Sa values points to a more smoothed surface structure, which is interpreted as a lack of repeatability in the focal positioning of the objective on the examined surface. An independent indicator is the Ssk parameter, interpreted according to the criteria discussed at the beginning of this section. Its overall analysis reveals a pronounced decrease in Ssk for the mean data in the sequential measurement mode, which indicates a reduction (smoothing) of the number of high peaks and–similarly to Sq and Sa – confirms the scatter in the focal positioning of the microscope objective. In addition, the analysis of difference surfaces can be used as an independent quantitative parameter describing the degree of texture similarity.

Another independent indicator of surface similarity is the fraction of NM points. This parameter should be interpreted as follows: the higher the value of the NM coefficient after surface differencing, the higher the accuracy of objective positioning and the greater the mutual similarity of the textures. Analysis of the distribution of remaining points after computing the difference surface also makes it possible to estimate the objective positioning error. In the present case, the maximum relative error determined on the basis of direct measurement of the distance between the remaining points was 51 μm. The core part of the analysis is the comparison of data for the reference surface and the surface measured sequentially. Figure 3 presents the differential surface data together with the visualization of the remaining points.

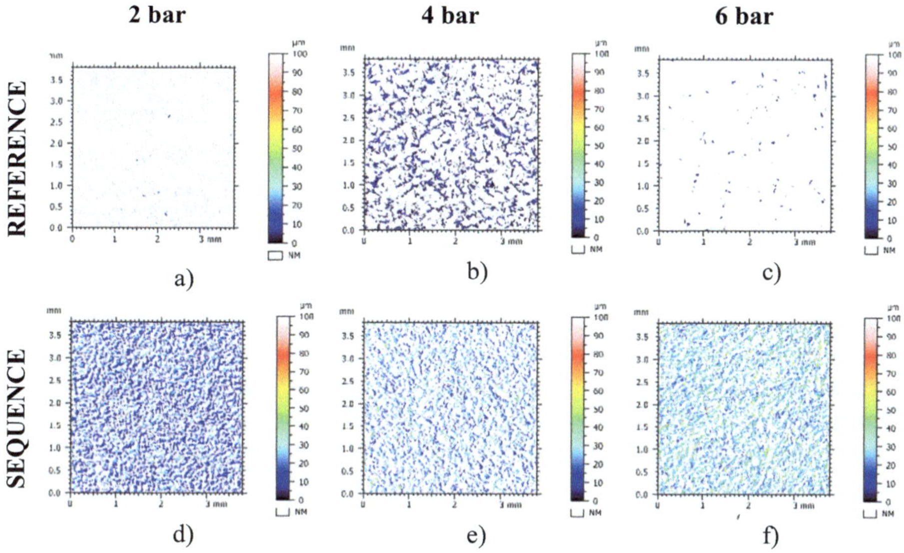

Fig. 3. Visualization of remaining points (differentiating areas) – a greater surface coverage and a higher peak height value = deterioration of the optical system's positioning accuracy

Figure 3 compares the data obtained for the considered measurement strategies. In the strategy where the objective displacement was limited (reference measurement), the differences between the same areas are small, which indicates good mutual matching of the surfaces. This is confirmed by the high number of NM pointsFthe more such points, the better the matching and the higher the measurement repeatability. Figure 3a–c show the results for anisotropic surfaces with different spatial parameters; in each of these cases, the differential points constitute only a small part of the analysed area.

A different situation is observed in the sequential variant (Fig. 3d–f), where the number of points remaining after mutual surface subtraction (absolute value) increases markedly. As a consequence, the NM coefficient takes lower values, which indicates poorer surface matching and lower repeatability of focal positioning. The numerical values for the analysed cases are presented in the form of a plot in Fig. 4.

The graph (Fig. 4) illustrates the differences between the measurement strategies in terms of the percentage fill of the measured surfaces. When analysing the data corresponding to the visualizations in Fig. 3, it is readily apparent that, in the case of the sequential measurement, the number of points remaining after mutual subtraction is clearly higher. This difference is visible for each of the analysed surfaces, although it

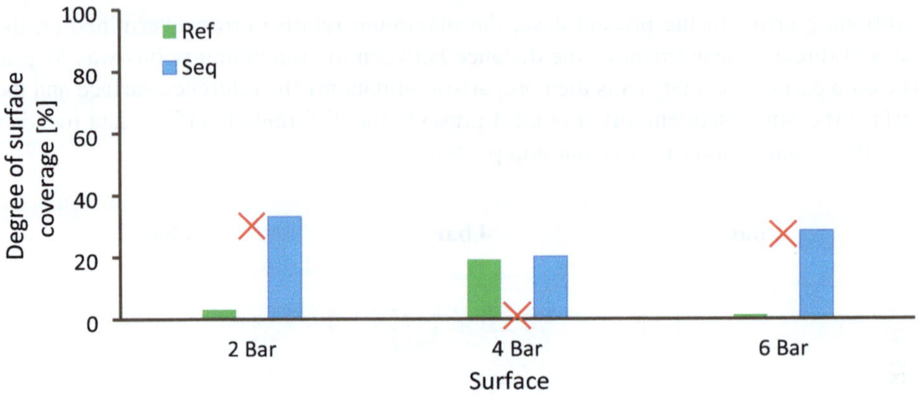

Fig. 4. Summary of remaining points from surface differencing for the investigated measurement strategies.

is least pronounced for the surface textured at a pressure of 4 bar. A plausible explanation is the binary interpretation of point presence – based solely on whether a point exists, without accounting for its amplitude. Even a point with a value close to ≈ 0 is still treated as an existing point. It is worth emphasizing that, for the surface considered in the 4 bar variant, the amplitude values after differencing are close to 0. This form of data presentation is particularly useful for qualitative comparisons.

4 Summary and Conclusions

The conducted investigations and the obtained results made it possible to formulate conclusions summarizing the metrological and functional analysis of the examined surfaces. The most important of these are presented below:

- In the reference configuration, the FV technique provides high repeatability of topography measurements. For the reference measurements, with minimal displacement of the objective, very small differences in the Sq and Sa parameters (approximately 1.3% and 0.4%, respectively) were obtained, which confirms the high stability of the focal positioning of the FV microscope objective.
- Long axial displacements of the objective significantly deteriorate the repeatability of positioning and the reproduction of the topography. In the sequential mode, with large displacements of the objective, the differences in the Sq and Sa parameters increase to approximately 22–23%, and the surface structure becomes more "smoothed", which indicates a lack of repeatability in focus positioning.
- The Sa, Sq and Ssk parameters effectively describe the influence of the measurement strategy on the reproduction of the surface structure. The observed decrease in Ssk values in sequential measurements indicates a reduction in the number of high peaks and a change in the character of the structure, which – similarly to the changes in Sa and Sq – confirms the spread in beam/objective positioning.

- The NM parameter and the analysis of difference surfaces constitute sensitive indicators of the quality of texture matching. A high value of the share of NM points after surface differencing correlates with good agreement of the topographies and high measurement repeatability, whereas a decrease in NM in the sequential variant reflects a deterioration of the matching and a larger positioning error.
- The analysis of points remaining after differencing enables a quantitative assessment of the positioning error. Based on the distances between the residual points, the maximum relative positioning error of the objective was estimated to be on the order of 51 μm, which demonstrates that the proposed methodology can be used for quantitative estimation of the uncertainty component associated with the mechanics of the system.
- The choice of scanning strategy is crucial for multi–repeat measurements of surface textures. The results indicate that in applications requiring high repeatability (quality control, process validation, comparative studies), strategies with limited axial displacement of the objective should be preferred, or the influence of long displacements should be compensated for in analytical procedures.

The proposed methodology can serve as a universal tool for the validation of non–contact measurement systems. The adopted analysis scheme (3D parameters, Ssk, NM, difference surfaces, correlation with positioning) can be adapted to other measurement systems and various types of textures, supporting the design of inspection procedures in surface engineering and the estimation of uncertainty under industrial and laboratory conditions. The applicability of the proposed method enables its use in other metrological instruments, particularly in systems where multiple measurement areas can be defined (CNC operating mode), such as confocal profilometry and white-light interferometry (WLI).

5 Funding

This scientific work was funded from the state budget under the program of the Minister of Science entitled "Polish Metrology II" (Poland), project number PM-II/SP/0038/2024/02, funding amount 999,530.40 PLN, total project value 999,530.40 PLN.

References

1. Maruda, R.W. et al.: Evaluation of turning with different cooling–lubricating techniques in terms of surface integrity and tribologic properties. Tribol. Int. **148**(106334), 1–14 (2020). https://doi.org/10.1016/j.triboint.2020.106334
2. Fetecau, C., Stan, F.: Study of cutting force and surface roughness in the turning of polytetrafluoroethylene composites with a polycrystalline diamond tool. Measurement. **45**, 1367–1379 (2012). https://doi.org/10.1016/j.measurement.2012.03.030
3. Pawlus, P., Reizer, R., Wieczorowski, M.: Functional importance of surface texture parameters. Materials. **14**(5326), 1–29 (2021). https://doi.org/10.3390/ma14185326
4. Brown, C.A. et al.: Multiscale analyses and characterizations of surface topographies. CIRP Ann. Manuf. Technol. **67**, 839–862 (2018). https://doi.org/10.1016/j.cirp.2018.06.001

5. Mital, G., Spišák, E.: 3D measurement of geometric characteristics of the surface. Int. J. Eng. Sci. (IJES). **11**(2), 56–61 (2022). https://doi.org/10.9790/1813-1102015661
6. Havelková, M., Hiklová, H.: Several examples of using contact profilometer for optical surface mapping. In: Proc. SPIE 8697, 18th Czech–Polish–Slovak Optical Conference on Wave and Quantum Aspects of Contemporary Optics, vol. 8697, p. 86972G (2012). https://doi.org/10.1117/12.2010362
7. Caja García, J., Sanz Lobera, A., Maresca, P., Fernández Pareja, T., Wang, C.: Some considerations about the use of contact and confocal microscopy methods in surface texture measurement. Materials. **11**(1484), 1–16 (2018). https://doi.org/10.3390/ma11081484
8. Walczak, D., Krolczyk, J.B., Chudy, R., Gupta, M.K., Pruncu, C., Krolczyk, G.M.: Role of optical measurement systems in analysing the surface topography of an industry standard component. Optik. **283**(170919), 1–14 (2023). https://doi.org/10.1016/j.ijleo.2023.170919
9. Aulbach, L., Salazar Bloise, F., Lu, M., Koch, A.W.: Non–contact surface roughness measurement by implementation of a spatial light modulator. Sensors. **17**(596), 1–15 (2017). https://doi.org/10.3390/s17030596
10. Gogolewski, D.: Evaluation of repeatability of topographic measurements based on multiscale analysis. Manuf. Lett. **45**, 17–20 (2025). https://doi.org/10.1016/j.mfglet.2025.06.202
11. Zanobini, A., Sereni, B., Catelani, M., Ciani, L.: Repeatability and reproducibility techniques for the analysis of measurement systems. Measurement. **86**, 125–132 (2016). https://doi.org/10.1016/j.measurement.2016.02.041
12. Wieczorowski, M., Cellary, A., Majchrowski, R.: The analysis of credibility and reproducibility of surface roughness measurement results. Wear. **269**, 480–484 (2010). https://doi.org/10.1016/j.wear.2010.05.003
13. Kuo, C.C., Huang, P.J.: Repeatability and reproducibility study of thin film optical measurement system. Optik. **124**, 3489–3493 (2013). https://doi.org/10.1016/j.ijleo.2012.10.015
14. Razumić, A., Runje, B., Keran, Z., Trzun, Z., Pugar, D.: Reproducibility of areal topography parameters obtained by atomic force microscope. Tehnički glasnik. **19**(1), 1–6 (2025). https://doi.org/10.31803/tg-20250324183037
15. Schmit, J., Novak, E.: Improved shape measurement in optical profiler. In: Proc. SPIE 5190, Recent Developments in Traceable Dimensional Measurements II, vol. 5190, pp. 193–202, San Diego, USA (2003). https://doi.org/10.1117/12.506946
16. Grochalski, K. et al.: Procedure for data preparation for the correct evaluation of the geometrical structure of the surface in functional analysis. Metrol. Hallmark. **8**(1), 1–13 (2024)
17. Pereira, M.D., Postolache, O., Girão, P.S.: Using neural network techniques in environmental sensing and measurement systems to compensate for the effects of influence quantities. IEEE Instrum. Meas. Mag. **17**(6), 26–56 (2014). https://doi.org/10.1109/MIM.2014.6968927
18. Leach, R.K., Giusca, C.L.: Determination of the metrological characteristics of optical surface topography measuring instruments. In: Proc. SPIE 8430, Optical Micro– and Nanometrology IV, p. 84300Q, Brussels, Belgium (2012). https://doi.org/10.1117/12.921452
19. Calvo, R., D'Amato, R., Gómez, E., Domingo, R.: Integration of error compensation of coordinate measuring machines into feature measurement: part II – experimental implementation. Sensors. **16**(1705), 2–18 (2016). https://doi.org/10.3390/s16101705
20. Leach, R.K., Giusca, C.L., Haitjema, H., Evans, C., Jiang, X.: Calibration and verification of areal surface texture measuring instruments. CIRP Ann. Manuf. Technol. **64**, 797–813 (2015). https://doi.org/10.1016/j.cirp.2015.05.010
21. Eifler, M., Seewig, J., Hering, J., von Freymann, G.: Calibration of z–axis linearity for arbitrary optical topography measuring instruments. In: Proc. SPIE 9525, Optical Measurement Systems for Industrial Inspection IX, vol. 9525, p. 952510, Munich, Germany (2015). https://doi.org/10.1117/12.2190737

22. Haitjema, H.: Calibration of displacement laser interferometer systems for industrial metrology. Sensors. **19**(4100), 2–21 (2019). https://doi.org/10.3390/s19194100
23. Vrba, I., Palencar, R., Hadzistevic, M., Strbac, B., Spasic–Jokic, V., Hodolic, J.: Different approaches in uncertainty evaluation for measurement of complex surfaces using coordinate measuring machine. Meas. Sci. Rev. **15**(3), 111–118 (2015). https://doi.org/10.1515/msr-2015-0017
24. Jamuła, B., Bazan, A., Magdziak, M.: Gage repeatability and reproducibility analysis of coordinate measurements of a cutting tool. Adv. Sci. Technol. Res. J. **17**(3), 133–144 (2023). https://doi.org/10.12913/22998624/165881
25. Peta, K., Love, G., Brown, C.A.: Comparing repeatability and reproducibility of topographic measurement types directly using linear regression analyses of measured heights. Precis. Eng. **88**, 192–203 (2024). https://doi.org/10.1016/j.precisioneng.2024.02.009
26. Danzl, R., Helmli, F., Scherer, S.: Focus variation – a robust technology for high resolution optical 3D surface metrology. Strojniški vestnik. J. Mech. Eng. **57**(3), 245–256 (2011). https://doi.org/10.5545/sv-jme.2010.1755
27. Grochalski, K., Wieczorowski, M., H'Roura, J., Le Goic, G.: The optical aspect of errors in measurements of surface asperities using the optical profilometry method. Front. Mech. Eng. **6**(12), 1–11 (2020). https://doi.org/10.3389/fmech.2020.00012
28. Harmatys, W., Gąska, A., Gąska, P., Gruza, M., Sładek, J.A.: Assessment of background illumination influence on accuracy of measurements performed on optical coordinate measuring machine eEuipped with video probe. Sensors. **21**(2509), 1–16 (2021). https://doi.org/10.3390/s21072509
29. Santolaria, J., Yagüe, J.A., Jiménez, R., Aguilar, J.J.: Calibration–based thermal error model for articulated arm coordinate measuring machines. Precis. Eng. **33**, 476–485 (2009). https://doi.org/10.1016/j.precisioneng.2009.01.002
30. Guillory, J., Truong, D., Wallerand, J.P., Alexandre, C.: Absolute multilateration–based coordinate measurement system using retroreflecting glass spheres. Precis. Eng. **73**, 214–227 (2022). https://doi.org/10.1016/j.precisioneng.2021.09.009
31. Jia, G., Cao, J., Zhang, X., Huang, N.: Ambient temperature–induced thermal error modelling for a special CMM at the workshop level based on the integrated temperature regression method. Int. J. Adv. Manuf. Technol. **121**, 5767–5778 (2022). https://doi.org/10.1007/s00170-022-09533-1
32. Šafarič, J., Dolšak, B., Klobučar, R., Ačko, B.: Analysis of thermal contribution to the measurement uncertainty in step gauge calibration. Precis. Eng. **66**, 52–61 (2020). https://doi.org/10.1016/j.precisioneng.2020.06.012
33. Harmatys, W., Gąska, A., Gąska, P., Gruza, M., Sładek, J.: Impact of warm–up period on optical coordinate measuring machine measurement accuracy. Measurement. **172**(108913), 1–20 (2021). https://doi.org/10.1016/j.measurement.2020.108913
34. Grochalski, K., Wieczorowski, M., Pawlus, P., H'Roura, J.: Thermal sources of errors in surface texture imaging. Materials. **13**(2337), 2–18 (2020). https://doi.org/10.3390/ma13102337
35. Nieslony, P. et al.: A novel use of non–contact methods in analyzing the physical and technical surface characteristics of different materials using electromagnetic measurement techniques. Measurement. **241**, 115672 (2025). https://doi.org/10.1016/j.measurement.2024.115672
36. Pertuz, S., Puig, D., Garcia, M.A.: Analysis of focus measure operators for shape–from–focus. Pattern Recogn. **46**, 1415–1432 (2013). https://doi.org/10.1016/j.patcog.2012.11.011
37. Zhou, Y., et al.: A survey of multi–focus image fusion methods. Appl. Sci. **12**, 6281 (2022). https://doi.org/10.3390/app12126281
38. Cepova, L., Kovacikova, A., Cep, R., Klaput, P., Mizera, O.: Measurement system analyses – gauge repeatability and reproducibility methods. Meas. Sci. Rev. **18**(1), 20–27 (2018). https://doi.org/10.1515/msr-2018-0004

39. Oliveira, J.E.F., Neto, F.R.A., Oliveira, S.T.: Suitability verification of a roughness measurement process using measurement system analysis and the cdl* capability coefficient. Acta IMEKO. **14**(3), 1–5 (2025). https://doi.org/10.21014/actaimeko.v14i3.1958
40. Al-Refaie, A., Bata, N.: Evaluating measurement and process capabilities by GR&R with four quality measures. Measurement. **43**(6), 842–851 (2010). https://doi.org/10.1016/j.measurement.2010.02.016

Selected Measurement Problems Related to Illumination Conditions in the Application of the Focus Variation (FV) Technique

Karol Grochalski(✉), Julia Radziejewska, Krzysztof Kupka, and Michał Wieczorowski

Faculty of Mechanical Engineering, Poznan University of Technology, Poznan, Poland
karol.grochalski@put.poznan.pl

Abstract. In this study, the influence of illumination disturbances on surface topography measurements performed using Focus Variation (FV) microscopy was investigated. Experiments were carried out on a sinusoidal roughness standard measured with an Alicona μCMM system under varying illumination conditions. The results show that seemingly negligible ceiling illumination significantly distorts the 2D microscopic image and markedly degrades the quality of the reconstructed 3D topography. Although the FV algorithms largely preserve the global directional character of the surface, the data exhibit increased noise and a pronounced rise in the number of non-measured points (NM), exceeding a tenfold increase compared to the reference conditions. The Ssk/Rsk and Sp/Rp parameters were found to be particularly sensitive to illumination disturbances, with Sp/Rp deviating by more than 46% under ceiling illumination. The combined illumination configuration mitigates some of these effects, improving the quality of the 3D data due to the dominant contribution of coaxial light. In summary, the study highlights that seemingly insignificant ambient light sources may constitute a significant and often overlooked source of measurement error in FV-based surface metrology.

Keywords: Focus Variation Microscopy (FV) · Surface Topography · Illumination Disturbances

1 Introduction

The development of technology worldwide is creating ever more opportunities for designing surface properties, as well as their machining and finishing. In addition to measurements of macro–geometry, optical measurement systems make it possible to measure the geometrical surface structure [1]. The need to assess the quality of engineering surfaces is well justified for industrial applications, which, as noted by Walczak et al. [2], determines the growing importance and applicability of optical measurement systems.

The increasing role of non–contact techniques results from their ability to perform fast and large–scale measurements, with particular emphasis on the geometrical structure

M. Hryb et al. (Eds.): MANUFACTURING 2026, LNME, pp. 49–60, 2026.
https://doi.org/10.1007/978-3-032-23038-6_4

of surfaces. Methods such as Focus Variation Microscopy (FV), described by Danzl et al. [3], which combines a small depth of field with vertical scanning along the Z–axis, have gained popularity due to their capability to measure surfaces with high slopes and varying roughness. It should be emphasized that even state–of–the–art measurement techniques are still not free from imperfections and errors affecting the quality of measurement data. Maculotti et al. indicate that optical instruments used for surface–structure measurements are susceptible to the occurrence of non–measured points and point disturbances in the form of "peaks", which result from complex interactions between the surface and the measurement system [1]. The complexity of this interaction is confirmed by Pahl et al. [4], who propose unified electromagnetic modelling for interferometric, confocal and FV microscopy in order to better predict systematic deviations arising from the wave nature of light. To maintain the accuracy of surface–structure measurements using optical methods, it is necessary to significantly limit the influence of the external environment. Thermal changes, shocks and vibrations, as well as lighting conditions may affect the reliability of the obtained results [5, 6]. Podulka et al. [5], in their work, highlight the issue of removing "peaks" from measurement data, noting that they affect the measured data, while their removal leads to a decrease in amplitude values describing the maximum height of selected surface–roughness parameters.

Measurement disturbances also have a thermal origin: thermal effects may cause a drift of the focal position due to heating of the immersion oil by the laser in confocal microscopy, thereby affecting measurement stability, as demonstrated by Matrecitos–Avila et al. [7]. As reported by Giusca et al. [8], in surface–structure measurements using Focus Variation Microscopy it is important to determine focus points using an appropriate level of contrast and structural illumination. This is critical for surfaces with poorly pronounced texture.

In response to this problem, Yuan et al. [9] propose the use of an active illumination mode employing a checkerboard pattern generated by a digital micromirror device (DMD), which introduces additional texture and improves the grayscale gradient necessary for focus evaluation. An alternative, passive approach is presented by Matsumoto et al. [10], who demonstrate that light polarization can be treated as a texture dependent on surface orientation. This enables accurate reconstruction of the three–dimensional surface shape at the microscale. Elango et al. [11], in their study, showed that optical parameters and image patterns depend on illumination conditions, including the angle of incidence, the distance between the light source and the sample, and the orientation of the sample with respect to the light. Tang et al. [12] presented the influence of measurement angles and light intensity on the reliability and accuracy of the data.

Since methods based on structured light rely on the principle of triangulation, they are inherently accompanied by the formation of shadows. Consequently, the measurement angle has a significant impact on the accuracy of obtained results in surface measurements and on the occurrence of errors, which may be observed regardless of the direction of light incidence. Complementing these observations, Nikolaev et al. [13] pointed out the sensitivity of Focus Variation Microscopy to surface inclination. At small angles of incidence, signal quality gradually deteriorates due to the superposition of specular and diffuse reflection components. Moreover, in optical microscopes, misalignment of the aperture stop may occur, leading to asymmetry of angular illumination, which results in lateral image shift and changes in focal length, as shown by Attota and Park [14].

Another important factor is illumination intensity. Addressing the issue of non–measured points in optical surface measurements, Pawlus et al. [15] investigated the influence of light intensity on the measurement results. The occurrence of non–measured points may be caused by many factors, including high surface steepness, strong light absorption or the reflectance coefficient. The study indicated that, as light intensity increases from a very low level, the proportion of non–measured points initially decreases, reaches a minimum, and then increases again. At too low illumination levels, non–measured points appeared mainly in the valleys of the surface, whereas excessive intensity generated them on peak regions. When the light intensity was increased, a general upward trend in height, hybrid and functional parameters was observed. Qi et al. [16] analysed the influence of image saturation on the accuracy of structured–light based methods. Their research showed that excessive intensity of reflected light, leading to detector overexposure, is a source of phase–calculation errors. To address pixel saturation on high-contrast surfaces, Hongzhi Jiang et al. [17] proposed the High Dynamic Range Focus Assessment (HDRFA) technique, which fuses a series of images acquired at different exposure times. This method was later extended by Chufan Jiang et al. [18] to real–time operation, where instead of changing exposure, the system uses complementary, inverted fringe patterns (shifted by 180°) to replace saturated data.

John et al. [19] indicated that, for surface roughness evaluation, illumination determining the quality of a given image is crucial, and this quality changes with varying illumination, i.e., reflections and lighting conditions. When illuminating the examined surface, a phenomenon may occur in which parts of the surface illuminate each other, leading to the formation of incorrect fringe patterns. This effect causes a marked increase in local noise in the measurements [20]. Furthermore, John et al. [21] showed that the resulting image, more specifically its texture, is affected by non–uniform illumination. To obtain consistency and, at the same time, accuracy in the roughness of a given surface, it is necessary to eliminate disturbing non–uniform illumination by applying, for example, filtering. Grochalski et al. [22] demonstrated that the type and direction of illumination significantly affect the obtained results in roughness measurements using Focus Variation Microscopy. The use of lens–generated illumination or collimated illumination emitted by a ring light makes it possible to obtain favourable conditions during surface measurements. Activating all sectors of the ring circuit provides a uniform distribution of light intensity. In turn, the use of directional illumination by activating selected ring sectors enables better observation of irregularities with random orientation and local peaks. Researchers also investigated the effect of a polarizing filter used to reduce surface reflections on surface topography parameters. The accuracy of the obtained surface–measurement results changes depending on the orientation of a sample with a regular, directional structure with respect to the polarizer axis. The polarization angle affected both the measured surface parameters and the cross–sectional profile projections. The highest values of Rp, Rz and Rt parameters were obtained when the polarizer axis was aligned parallel to the direction of grooves and scratches.

Independently, Harmatys et al. [23] emphasize that, in optical methods, measurement results are influenced by improper focus and the parameters of the applied illumination.

To investigate the effect of luminance, measurements were carried out using maximum magnification and different levels of illumination intensity. The analysis showed that increasing illumination intensity most strongly affects the measurement of the diameter of the measured elements.

Based on the literature review, it can be concluded that there is a lack of sufficient studies verifying the influence of illumination originating outside the measuring device itself, such as room lighting, sunlight or other sources not directly coordinated with the settings of the measuring apparatus. The aim of this work is to investigate how selected disturbances – including additional external light sources – affect the formation of measurement errors in optical surface–topography metrology. The motivation for undertaking this research was the observation of differences in the image and in the obtained results during surface measurements performed with the laboratory room lighting switched on and off.

2 Methodology

The topography of the analysed surfaces was captured using an Alicona μCMM optical measurement system operating on the principle of Focus Variation (FV) microscopy. The applied method enables non-contact acquisition of three-dimensional topographic data by combining imaging with a large depth of field and precise axial scanning. During the measurement, the objective (or specimen) is sequentially displaced along the Z-axis, and for each position the contrast distribution in the focal plane is recorded. By maximising a sharpness metric for individual pixels, the height of surface points can be determined and the three-dimensional structure of the surface can be reconstructed.

In contrast to classical stylus profilometry, the FV technique does not require physical contact between the probe tip and the examined object, which eliminates the risk of mechanical interference with the surface structure – both in the form of local deformations and artefacts resulting from the geometrical limitations of the stylus. The absence of mechanical interaction also prevents the "smoothing" of topography at the micro- and nanometric scale, which is typical of contact methods. As a result, the obtained height maps exhibit improved conformity with the true surface morphology, which is particularly important for the evaluation of roughness and surface texture parameters as well as for the analysis of functional properties such as load-bearing capacity, lubricant retention, or wear resistance.

Additionally, the use of a μCMM system integrating the FV technique with the functionality of a coordinate measuring machine makes it possible to link local topography measurements with macrogeometrical measurements. This approach promotes metrological consistency of the results, facilitates tracing of measurement uncertainty sources, and enables a more comprehensive characterisation of the examined components – from form and shape geometry through to detailed analysis of the surface texture. To verify measurement repeatability, each test was performed five times under stable thermal conditions. Each measurement took less than 10 min. This ensured that potential errors related to environmental disturbances were minimized.

For the purpose of parametric analysis of the surface under disturbed illumination conditions, a representative measurement area was selected on a sinusoidal roughness standard with nominal parameters Ra = 1 μm, Rsm = 100 μm. The measurement area was defined as a rectangle of 0.6 mm × 7.2 mm, with the larger dimension aligned with the periodic direction of the surface. Data acquisition was carried out at 20× magnification (objective 400WD19). The vertical (Z-axis) resolution was 30 nm, and the lateral resolution was 500 nm. The measurement range along the Z-axis was set to 100 μm.

The standard was mounted in a dedicated fixture attached to the tilting-rotary stage of the microscope. The measurement setup is shown in Fig. 1a. Figure 1b presents the ceiling lighting of the laboratory housing the Alicona μCMM. This lighting constitutes a potential source of errors in the imaging of the geometrical surface structure, particularly when measuring specimens with low amplitude values and high reflectivity.

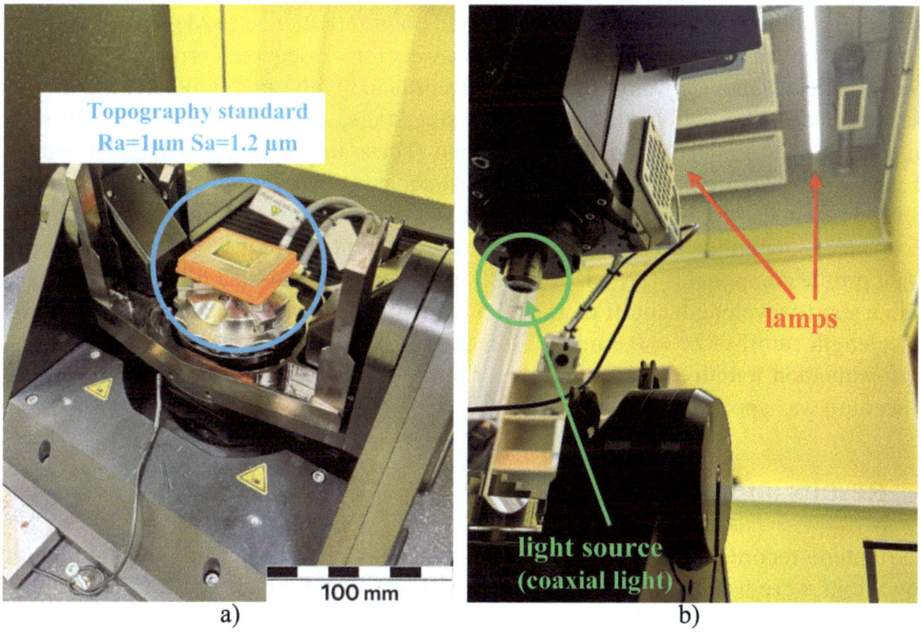

a) b)

Fig. 1. Measurement setup – focus variation microscope: a) measured object – roughness standard, b) laboratory lighting – potential source of disturbance (position of the lamps determines the nature of the resulting disturbances)

The measurement methodology involved assessing the influence of external illumination, such as ceiling lamps, on disturbances in the spatial reproduction of the surface topography. The verification procedure was divided into three main operations:

- measurement of the surface of the standard with the laboratory lighting and other potential disturbance sources switched off. This measurement is treated as the reference measurement. The illumination source was laser light with a wavelength of 530 nm (green) transmitted through the microscope objective (coaxial light), with an intensity of I = 10%, contrast C = 25%, and exposure time B = 500 μs,

- measurement of the surface of the standard with the coaxial light switched off; the specimen was illuminated exclusively by the ceiling lighting of the laboratory, whose parameters are not parametrically defined (the same lighting that illuminates the laboratory and may constitute a constant component of any measurement). In this variant, the contrast was kept unchanged at $C = 25\%$, whereas the exposure time was adjusted based on the illumination intensity histogram to obtain conditions as close as possible to the reference measurement. The resulting exposure time in this case was $B = 1.1$ ms,
- measurement of the surface with simultaneous illumination by the coaxial light ($I = 10\%$, contrast $C = 25\%$, exposure time $B = 500$ μs).

Based on the recorded data, the geometrical surface structure was analysed using MountainsMap® Premium 10.3 software. The software enabled the determination surface topography parameters, the non-measured points coefficient (NM), visualisation of the examined surfaces, and preparation of datasets for subsequent correlation analyses. The parameter analysis was conducted in accordance with, among others, the recommendations of ISO [24]. At the data-processing stage, standard procedures consistent with good metrological practice were applied. These included: form removal using a second-order polynomial fit, levelling by least-squares plane fitting, and spatial filtering to suppress micro-roughness, noise in the form of isolated spikes (overexposure artefacts), and other artefacts typical of optical measurement techniques. To separate the relevant profile components (primarily for 2D evaluation), a Gaussian filter was employed with a short-wavelength cutoff of $\lambda s = 2.5$ μm (S-filter for noise reduction) and a long-wavelength cutoff of $\lambda c = 0.8$ mm (L-filter/cutoff for roughness–waviness separation). The evaluation length also satisfied the criterion of $5 \times \lambda c$. The same computational procedure was applied for each experiment.

3 Results and Discussion

The textures recorded for the individual measurement variants under varying illumination conditions were used to assess the influence of additional, disturbing light sources on the acquired topography. Figure 2 juxtaposes two-dimensional images from the direct microscopic live view (2D) with three-dimensional topographic data (3D) reconstructed in the form of a point cloud. Figure 2a shows the reference measurement obtained using coaxial illumination (coaxial light). The remaining structures, measured in accordance with the procedure described in Sect. 2, are compared against these data. Particular attention should be paid to the direct microscopic image (2D – left grayscale panel): the sinusoidal structure exhibits pronounced directionality in the form of vertical lines and a uniform distribution of light intensity, with no visible overexposed regions or artefacts. The 3D reconstruction (right, colour panel) is characterised by good coverage with measurement points; the sinusoidal geometry of the standard is clearly resolved and does not exhibit distortions in the form of "peaks" or non-measured regions (NM points).

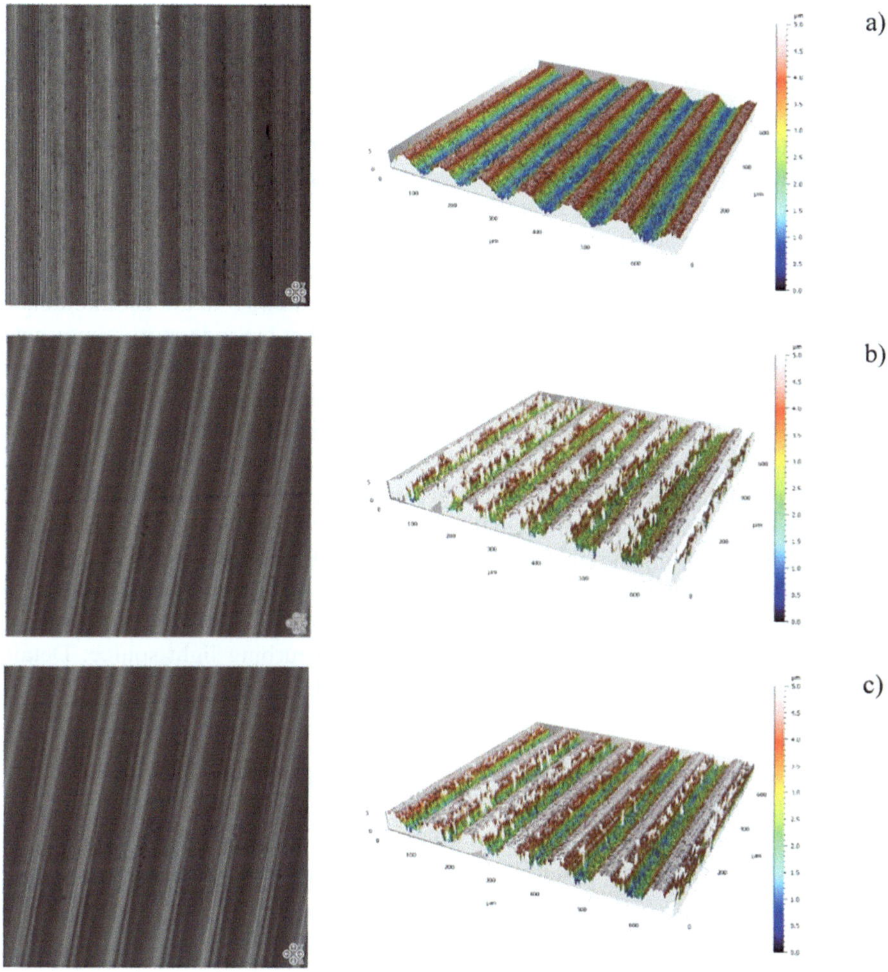

Fig. 2. View of the measurement standard surface under different lighting conditions (live microscope view and 3D surface topography visualization): a) coaxial light–reference, b) only lamps in laboratory, c) coaxial light + lamp

Highly informative observations can be made from the images shown in Fig. 2b–c. The surfaces were illuminated either by external lighting (the standard laboratory room illumination) or by a combination of this external lighting and coaxial illumination. Of particular interest is the case with ceiling-lamp illumination. Although such lighting is not intended for metrological applications, it still enabled three-dimensional measurement of the surface structure. In the configuration considered here, this type of illumination precludes precise adjustment of luminance (the intensity depends on distance, angle of incidence, source power, and other factors); nevertheless, the light is reflected from the measured surface, thereby determining the spatial distribution of intensity and focus. In the laboratory of the Division of Metrology and Measurement Systems at Poznan

University of Technology, two longitudinal ceiling lamps are installed above the measurement station. As a consequence, the resulting image is characterised by a "crossed" and "tilted" pattern of lines forming the geometry of the standard. The surface of the standard remains unchanged, but the microscopic image is distorted. The crossed and inclined pattern depends on the direction of light incidence. Despite the noticeable distortion of the surface in the 2D microscopic image, the topography is measured in a reasonably correct manner. The overall character of the structure is preserved, and the tilt of the vertical lines of the standard is not reflected in the reconstructed topography. This indicates that the depth-of-focus detection algorithms exhibit limited sensitivity to this type of distortion. However, a substantial increase in noise in the acquired data is evident. There are clearly visible regions (NM points) that are incorrectly represented, and the unique features of the standard are therefore not fully captured. The conclusion is that the general character of the surface (structural directionality) is preserved, but its quality is significantly degraded compared to the recommended illumination conditions.

For the surfaces shown in Fig. 2c, corresponding to the combination of coaxial microscope illumination and the external light source, the appearance of the surface in the microscopic image is almost identical to the previously discussed case. The characteristic tilting of the lines forming the standard is again visible. However, the quality of the three-dimensional data is markedly better. The number of disturbed regions is reduced, and the geometry of the standard remains interpretable. This improvement is attributed to the dominance of the coaxial illumination over the disturbing light source. Detailed information on the surface parameters is summarised in Table 1.

Table 1. Surface topography parameters for different measurement lights conditions

	coaxial light (Ref)		lamps		coaxial + lamps	
	Surface	Profile	Surface	Profile	Surface	Profile
Sa / Ra [μm]	1.18	1.16	1.40	1.37	1.39	1.38
Sq / Rq [μm]	1.35	1.32	1.66	1.64	1.65	1.63
Sp / Rp [μm]	2.35	2.36	3.44	3.32	3.42	3.43
Sz / Rz [μm]	4.7	4.71	6.54	6.31	6.52	6.57
Sv / Rv [μm]	2.35	2.35	3.10	2.99	3.10	3.14
Ssk / Rsk [−]	0.003	0.022	0.144	0.137	0.133	0.132
Sku / Rku [−]	1.74	1.75	2.16	2.17	2.14	2.14
NM [%]	0.944		11.4		11.6	

The parameters Ssk / Rsk. which clearly describe the asymmetry of the surface height distribution. are also strongly correlated with the disturbing light source – indicating. in simplified terms. whether the surface is "more peak-dominated" or "more valley-dominated". By definition. Ssk / Rsk $>$ 0 (positive) means that the examined surface exhibits a predominance of peaks over valleys. which is in strict accordance with the images shown in Fig. 2b.

When considering all parameters listed in Table 1. particular attention should be paid to the coefficient of NM points. which specifies the number or proportion of nodes in the measurement grid for which the measurement system failed to register a valid height value. For illumination variants other than coaxial. the differences in the NM coefficient exceed an order of magnitude (more than tenfold).

The tabulated comparison of parameters indicates substantial differences between the individual spatial parameters. A graphical representation of the spatial parameters is presented in the form of a graph showing percentage deviations relative to the reference data.

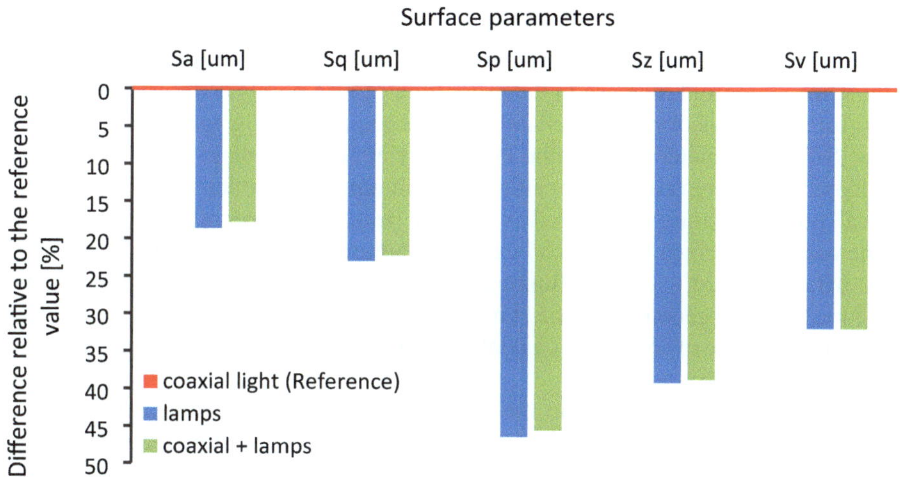

Fig. 3. Summary of selected surface texture parameters – percentage differences relative to the reference measurement.

The graph (Fig. 3) shows the percentage deviation of the surface parameters recorded under ceiling-lamp illumination and combined illumination. Each parameter under consideration exhibits the highest relative deviation in the presence of ceiling lighting. The largest discrepancies with respect to the reference measurement were found for the Sp / Rp parameter. exceeding 46% for ceiling-lamp illumination – the potential source of disturbance – which may indicate localised noise in the form of isolated peaks. It should be borne in mind that this parameter is highly sensitive to various types of interference and fluctuations.

4 Summary and Conclusions

Based on the conducted experiments and the analysis of the obtained results. it was possible to formulate conclusions regarding the influence of illumination conditions on the reliability of surface topography measurements performed using Focus Variation (FV) microscopy. The most important metrological and practical observations are summarised below; they may serve as a reference both for the interpretation of measurement results and for the design of measurement procedures that account for environmental disturbances.

- External ceiling illumination significantly perturbs the FV topography acquisition process. inducing distortions in the 2D images (cross-like patterns and tilted lines) and leading to a pronounced degradation in the quality of the reconstructed surface structure.
- Although the 2D microscopic image is visibly distorted under external illumination. the FV algorithms preserve the overall directional character of the surface; however. this occurs at the expense of a marked increase in measurement noise and a loss of distinctive features of the reference standard.
- The Ssk/Rsk parameters and the NM point coefficient are highly sensitive to illumination disturbances. For configurations other than the reference coaxial illumination. more than a tenfold increase in the proportion of NM points is observed. indicating a substantial reduction in the reliability of the acquired topographic data.
- The Sz/Rz parameter exhibits the largest deviation (exceeding 46% relative to the reference measurement) under ceiling-lamp illumination. which can be attributed to local. point-like noise manifested as artificial peaks in the reconstructed topography.
- The use of combined coaxial and external illumination improves the quality of the 3D data compared to ceiling illumination alone: the number of disturbed regions is reduced and the geometry of the standard remains clearly interpretable. confirming the dominant role of controlled coaxial illumination in achieving accurate topography reconstruction.
- Based on the observations and the acquired data, it is recommended to perform measurements using Focus Variation technique in dimly lit or completely dark rooms. This is particularly important when measuring highly reflective surfaces.

Although these data represent a specific case. they provide an important indication of potentially overlooked sources of error in surface measurement using the Focus Variation technique. This observation forms a foundation for further research aimed at identifying broadly understood environmental sources of measurement error.

Funding. This research was funded by the Ministry of Science and Higher Education grant No. 0614/SBAD/1586.

References

1. Maculottia, G., Gentaa, G., Quagliottib, D., Hansenb, H.N., Galettoa, M.: Comparison of methods for management of measurement errors in surface topography measurements. Procedia CIRP. **118**, 1084–1089 (2023). https://doi.org/10.1016/j.procir.2023.06.186
2. Walczak, D., Krolczyk, J.B., Chudy, R., Gupta, M.K., Pruncu, C., Krolczyk, G.M.: Role of optical measurement systems in analysing the surface topography of an industry standard component. Optik. **283**(170919), 1–14 (2023). https://doi.org/10.1016/j.ijleo.2023.170919
3. Danzl, R., Helmli, F., Scherer, S.: Focus variation – A new technology for high resolution optical 3d surface metrology. In: The 10th International Conference of the Slovenian Society for Non–Destructive Testing, pp. 481–491 (2009)

4. Pahl, T. et al.: Electromagnetic modeling of interference. confocal. and focus variation microscopy. Adv. Photonics Nexus. **3**(1) 016013, 1–13 (2024). https://doi.org/10.1117/1.APN.3.1.016013
5. Podulka, P., Pawlus, P., Dobrzański, P., Lenart, A.: Spikes removal in surface measurement. 14th Int. Conf. Metrol. Prop. Eng. Surf. J. Phys.: Conf. Ser. **483**, 1–9 (2014). https://doi.org/10.1088/1742-6596/483/1/012025
6. Podulka, P.: The effect of spikes occurrence on surface texture parameter assessments. IOP Conf. Ser.: Mater. Sci. Eng. **473**, 012039 (2019). https://doi.org/10.1088/1757-899X/473/1/012039
7. Matrecitos–Avila, M., Avila, R., Pimentel–Domínguez, R., Cuevas, S., Tamariz, E., Loza–Alvarez, P.: Focus variation due to near infrared laser in a confocal microscope. Microsc. Res. Tech. **85**, 3431–3438 (2022). https://doi.org/10.1002/jemt.24198
8. Giusca, C.L., Claverley, J.D., Sun, W., Leach, R.K., Helmli, F., Chavigner, M.P.J.: Practical estimation of measurement noise and flatness deviation on focus variation microscopes. CIRP Ann. Manuf. Technol. **63**(1), 545–548 (2014). https://doi.org/10.1016/j.cirp.2014.03.086
9. Yuan, L., Guo, T.: Active illumination mode with checkerboard pattern in focus variation microscopy: analysis and application. Opt. Lasers Eng. **184**, 108584 (2025). https://doi.org/10.1016/j.optlaseng.2024.108584
10. Matsumoto, R., Kawahara, R.: Polarization as texture: microscale 3D shape from polarized light focus. In: IEEE/CVF Winter Conference on Applications of Computer Vision, pp. 515–524 (2025). https://doi.org/10.1109/WACV61041.2025.00060
11. Elango, V., Karunamoorthy, L.: Effect of lighting conditions in the study of surface roughness by machine vision – an experimental design approach. Int. J. Adv. Manuf. Technol. **37**, 92–103 (2008). https://doi.org/10.1007/s00170-007-0942y
12. Tang, Y., Zhang, X., Liang, T.: Study on the error of the three–dimensional structure light measurement based on triangulation principle. Appl. Mech. Mater. **201–202**, 251–255 (2012). https://doi.org/10.4028/www.scientific.net/AMM.201-202.251
13. Nikolaev, N., Petzing, J., Coupland, J.: Focus variation microscope: linear theory and surface tilt sensitivity. Appl. Opt. **55**(13), 3555–3565 (2016). https://doi.org/10.1364/AO.55.003555
14. Attota, R.K., Park, H.: Optical microscope illumination analysis using through-focus scanning optical microscopy. Opt. Lett. **42**, 2306–2309 (2017). https://doi.org/10.1364/OL.42.002306
15. Pawlus, P., Reizer, R., Wieczorowski, M.: Problem of non–measured points in surface texture measurements. Metrol. Meas. Syst. **24**(3), 525–536 (2017). https://doi.org/10.1515/mms-2017-0046
16. Qi, Z., Wang, Z., Huang, J., Xing, C., Gao, J.: Error of image saturation in the structured–light method. Appl. Opt. **57**(1), 181–188 (2017). https://doi.org/10.1364/AO.57.00A181
17. Jiang, H., Zhao, H., Li, X.: High dynamic range fringe acquisition: A novel 3–D scanning technique for high–reflective surfaces. Opt. Lasers Eng. **50**(10), 1484–1493 (2012). https://doi.org/10.1016/j.optlaseng.2011.11.021
18. Jiang, C., Bell, T., Zhang, S.: High dynamic range real–time 3D shape measurement. Opt. Express. **24**(7), 7337–7346 (2016). https://doi.org/10.1364/OE.24.007337
19. John, J.G., Arunachalam, N.: Improved surface roughness evaluation of ground components using illumination compensated image: a machine vision approach. In: Proceedings of the ASME 2016 International Mechanical Engineering Congress and Exposition, pp. 1–11, Arizona. USA (2016). https://doi.org/10.1115/IMECE2016-66514
20. Munkelt, C., Kühmstedt, P., Heinze, M., Suesse, H., Notnia, G.: How to detect object–caused illumination effects in 3D fringe projection. Opt. Meas. Syst. Ind. Insp. **IV 5856**, 632–639 (2005). https://doi.org/10.1117/12.612895
21. John, J.G., Arunachalam, N.: Illumination compensated images for surface roughness evaluation using machine vision in grinding process. Procedia Manuf. **34**, 969–977 (2019). https://doi.org/10.1016/j.promfg.2019.06.099

22. Grochalski, K., Wieczorowski, M., H'Roura, J., Le Goic, G.: The optical aspect of errors in measurements of surface asperities using the optical profilometry method. Front. Mech. Eng. **6**(12), 1–11 (2020). https://doi.org/10.3389/fmech.2020.00012
23. Harmatys, W., Gąska, A., Gąska, P., Gruza, M., Składek, J.A.: Assessment of background illumination influence on accuracy of measurements performed on optical coordinate measuring machine equipped with video probe. Sensors. **21**(2509), 2–16 (2021). https://doi.org/10.3390/s21072509
24. ISO: ISO 25178-1: Surface texture: Areal — Part 1: Indication of surface texture. International Organization for Standardization (2016)

Analysis of the Impact of Resolution and Sampling Density on the Accuracy and Repeatability of Optical Coordinate Measurement

Ksenia Ostrowska(✉), Danuta Owczarek, Katarzyna Składanowska, Izabela Sanetra, and Tobiasz Klimczak

Cracow University of Technology, Kraków 31-155, Poland
ksenia.ostrowska@pk.edu.pl

Abstract. The development of measurement methods in the era of Industry 4.0 and 5.0 has made 3D scanning one of the most universal methods for acquiring geometric data. In the literature, the term resolution is used ambiguously, referring to detector resolution, structural resolution, or sampling resolution. In this paper, sampling resolution is defined as the distance between adjacent measurement points, which determines the density of the acquired data. In industrial practice, this parameter is often selected intuitively, which can lead either to redundant data generation or to a loss of geometric detail. The aim of this work is to evaluate the impact of the specified sampling resolution and the resulting data density on the accuracy, repeatability, and quality of geometric feature mapping. The tests were conducted on a ball-bar standard using a mobile metrological scanner based on laser triangulation. The geometric deviation results were compared with reference values obtained from a coordinate measuring machine. The analysis revealed a nonlinear relationship between sampling density, scanning time, and reconstruction accuracy, as well as the presence of an information-saturation threshold beyond which increasing the resolution no longer improves the results. The research was supplemented with tests aimed at assessing the system's ability to map small-diameter holes. The work addresses the problem of terminological ambiguity. The results contribute to the development of an Expert system with machine-learning components that enables rapid estimation of measurement uncertainty based on archival data. The tools being developed support advanced decision-making in selecting measurement systems for specific industrial applications.

Keywords: Optical Coordinate System · Triangulation · Resolution · Sampling Density · 3D Scanner · Multi-Criteria Matrix

1 Introduction

3D scanning is a key tool in coordinate metrology, enabling fast and non-destructive acquisition of information about the geometry of objects. The development of systems based on laser triangulation and structured light in the era of Industry 4.0 and 5.0 [1, 2]

M. Hryb et al. (Eds.): MANUFACTURING 2026, LNME, pp. 61–75, 2026.
https://doi.org/10.1007/978-3-032-23038-6_5

enables precise measurements to be taken directly in the production environment [3, 4]. With the growing popularity of 3D scanners, there is a need to select the right key data acquisition parameters.

The challenges associated with the proper characterization of optical measurement systems have long been present in scientific literature. As early as 2009, the authors of [5] drew attention to the problem of inconsistent manufacturer specifications for 3D scanners and cameras, which significantly hinders the comparison of devices and their objective evaluation. The aim of that research was to develop a characterization process that would allow the capabilities of devices to be determined in terms of spatial resolution and accuracy (including shape, orientation, and surface waviness. In subsequent years, standardization activities were taken in response to these challenges. In 2015, in article [6], the authors presented progress in the standardization of methods for assessing accuracy, resolution, and repeatability within international standardization organizations (ISO, VDI/VDE). The need for proper metrological characterization of 3D sensors is also emphasized in the context of automation, where the depth resolution of various 3D sensors was analyzed to ensure precision in production systems [7].

An important parameter that directly affects the quality of the results obtained is the sampling resolution, which defines the specified distance between adjacent points recorded by the optical sensor. This resolution directly determines the sampling density, i.e., the number of points collected per unit area. The quality of geometric detail reproduction, as well as the size of the resulting files and data processing time, depend on the correct resolution setting. Despite the important role of data density for the final metrological quality, this issue is still treated in a general manner in the literature and international metrological standards (including ISO 10360-8 [8] and ISO 10360-13 [9]) in a general manner. These standards do not indicate a methodology for quantitative assessment and optimization of the impact of data density changes on the accuracy and repeatability of measurement results. As a result, in industrial practice, this parameter is often selected intuitively, which can lead to both inaccurate geometry mapping and loss of key details, as well as unnecessary overloading of the measurement system with excess data.

In view of the above, it is necessary to conduct research that will quantitatively determine the impact of resolution/sampling density on the quality of metrological measurements. The article discusses the impact of the specified sampling resolution (and thus the resulting data density) on the accuracy, repeatability, and geometric fidelity of features, as well as the identification of a nonlinear relationship between point density and measurement error, in order to determine the optimal density threshold for minimizing errors and acquisition time. The tests were performed using a mobile metrology scanner based on laser triangulation with a blue light projector, focusing on the measurement of a ball-bar standard, which enabled a precise comparison of dimensional deviations (diameter and distance between balls) with reference values determined on a CMM system. The research weas supplemented with tests using an articulated arm coordinate measuring machine aimed at assessing the system's ability to map small diameter holes [10, 11].

This publication is part of a broader research project whose long-term goal is to develop a multi-criteria matrix for selecting optical systems for specific measurement tasks. Ultimately, this matrix is to be integrated with PMI (Product and Manufacturing Information) systems, which will enable the automatic selection of optimal scanning parameters based on the geometric characteristics and tolerance requirements of the component. There is a clear trend in the development of metrology tools towards digital twins [12, 13] or virtual machines [14, 15]. The analysis of resolution as a factor determining the measurement result is part of the work within the LIDER15/0306/2024 project "An expert system based on machine learning as a new tool for modeling the uncertainty of coordinate measurement" supported by the National Centre for Research and Development under the LIDER program. The aim of the project is to develop a tool for quick and simple estimation of measurement uncertainty based on archival data, prior to the actual measurement, which will then be used in the tool selection system for a given task. Subsequent stages of the work will include extending the research to analyze the impact of structural resolution and comparing different optical data acquisition technologies.

2 Sampling Density/Resolution in 3D Scanning – Research Problem

2.1 The Concept of Resolution in 3D Scanning Systems

In the literature on optical metrology, the term resolution is used in many contexts. This ambiguity often leads to misinterpretations and makes it difficult to compare measurement systems. Device manufacturers often use this term in a general and confusing way, blurring the lines between the hardware capability of the sensor and the configuration parameter controlled by the operator. As a result, it is necessary to precisely distinguish and define the different types of resolution to ensure the reliability and comparability of metrological data. In triangulation-based systems, there are several independent types of resolution:

- Detector matrix resolution: refers to the number of pixels in the matrix on which the laser light reflection is recorded. It determines the minimum laser line shift that allows two points in the image to be distinguished. Higher detector resolution increases the theoretical ability of the system to distinguish fine details.
- Structural resolution: this is the minimum size of a geometric detail that the system is able to reproduce under the assumed measurement conditions. Structural resolution depends on the optics, operating point, laser power, filtration settings, triangulation geometry, and reconstruction algorithms. It is not a parameter that can be changed by the user.
- Sampling resolution: unlike the previous two, this is a configuration parameter controlled by the scanner operator. It specifies the minimum spatial distance between successive measurement points recorded by the device. This parameter directly determines the sampling density of the point cloud, i.e., the number of points collected per unit area.

The impact of sampling resolution (and thus the resulting sampling density) on measurement accuracy, repeatability, and reproducibility is the main subject of analysis in this paper. Controlling this parameter allows for a quantitative assessment of the trade-off between data quality and acquisition efficiency.

2.2 Research Station and Methodology

The research was planned in such a way as to enable a quantitative assessment of the impact of sampling resolution and thus the resulting specified sampling density on the quality of geometric feature mapping in measurements performed with a mobile 3D scanner and an articulated arm coordinate measuring machine with a laser head. For the purposes of analysis, the sampling resolution was taken as an independent variable, which directly translated into the density of points used in the fitting process. Both local features, such as ball diameters, and global features, i.e., the length of the ball-bar defined as the distance between the centers of the balls, as well as the diameter and depth of the hole, were analyzed. The measurement system and test procedure used were selected to ensure repeatable experimental conditions and the possibility of direct comparison of the results with reference values. The research, as well as the calibration of the systems, was carried out under controlled conditions in an accredited calibration laboratory, in accordance with the assumptions of the ISO standards chain, concerning environmental conditions and data processing algorithms consistent with the conditions and assumptions for the calibration of such systems [8, 9].

2.2.1 Research Assumptions and Objectives

The main objective of the research was to determine the extent to which a change in sampling density, resulting from manipulation of the sampling resolution in a triangulation system, affects the reproduction of basic geometric elements. The key issue was to determine whether there is a range of resolution that can be considered optimal in terms of measurement accuracy and stability (repeatability). It was assumed that sampling resolution, as a parameter available to the operator and changed during recording, leads to a nonlinear relationship between mapping quality. It was assumed that too low data density would result in reduced accuracy and repeatability, while excessively high density would only cause a marginal improvement in results, disproportionate to the increase in acquisition time and data volume.

2.2.2 Measurement Procedure

The ball bar measurement (reference values: ball diameters, distance between their centers) (See Fig. 1a) was performed at five different sampling resolutions: 0.1 mm, 0.5 mm, 1.0 mm, 2.0 mm, and 4.0 mm, in accordance with the range of available scanner settings. For each configuration, a series of repetitions were performed to assess the stability and variability of the results.

The reference values for the diameters of the balls and the distances between their centers were determined using a reference coordinate measuring machine (CMM), which enabled a comparison based on absolute difference and deviation analysis. Supplementary tests included tests using an articulated arm coordinate measuring machine equipped with a laser head to assess the system's ability to reproduce the features of holes on a dedicated test plate printed using SLA technology (diameter, depth) with small diameters (see Fig. 1b) at variable sampling resolutions of 0.05 mm, 0.08 mm, 0.1 mm, 0.2 mm, 0.3 mm, 0.4 mm, and 0.5 mm.

Fig. 1. a) Ball bar artefact used in the study, and b) SLA-manufactured plate with cylindrical and square holes.

The methodology used is comparative and fully controlled, which allows for the unambiguous attribution of changes in the results to the influence of the tested parameter – sampling resolution.

2.2.3 Equipment Used

A Creaform HandySCAN BLACK + Elite mobile metrology scanner, based on laser triangulation using blue laser projections, was used for data acquisition. Optoelectronic measurement systems using laser triangulation are currently one of the key tools in coordinate metrology, especially in the measurement of objects with complex shapes, delicate surfaces, and requiring short data acquisition times. The principle of operation of triangulation-based systems is to project a laser beam or line onto the surface of an object, the position of which is then recorded by a CCD/CMOS camera at a specific angle to the projector. By knowing the geometry of the system and processing the position of the visible laser line on the matrix, it is possible to determine the three-dimensional position of points on the surface of the measured object (Fig. 2).

Fig. 2. Schematic representation of laser triangulation [16] and the Creaform HandySCAN BLACK + Elite scanner used in the study.

This technology is distinguished by its high recording speed, the ability to obtain dense point clouds, and a scanning process that is relatively insensitive to the operator. However, it should be emphasized that, unlike classic coordinate measuring machines (CMMs), the accuracy of triangulation systems depends to a large extent on surface quality, lighting conditions, reconstruction algorithms, and crucially for this work acquisition parameters, including sampling resolution. Key features of the device:

- measurement accuracy: ±0.025 mm,
- adjustable sampling resolution (manual selection of resolution up to 0.1 mm),
- acquisition speed: 1,800,000 points/s,
- tracking system with reference markers,
- projection: 15 laser crosses (enabling data acquisition even from objects with complex geometry).

Supplementary tests were performed using the ROMER Absolute Arm RA-7320 SI AACMM with a laser triangulation probe, with an accuracy of approximately ±38 μm.

2.2.4 Software Used

Two specialized software environments were used in the data acquisition and processing process, dedicated to 3D scanning and advanced metrological analysis, respectively:

VXelements (Creaform), which served as a working platform for configuring scanning parameters, calibrating the measuring head, acquiring measurement data, pre-cleaning the point cloud, and generating triangular STL meshes for each sampling resolution. All STL meshes were exported without additional smoothing to preserve the raw surface characteristics resulting from the selected sampling resolution.

PolyWorks|Inspector (InnovMetric), which was used to fit nominal geometries (spheres, axes, planes), determine the distances between the centers of the ball bar spheres, diameter and depth values, analyzing geometric deviations and tolerances, generating 3D deviation maps, evaluating surface reconstruction stability, and comparing results with CMM reference values. PolyWorks enables objective and repeatable geometry fitting using the least-squares method, making it an industry standard for 3D scan data analysis.

3 Results

This chapter presents the results of ball bar measurements performed with a mobile metrology-grade 3D scanner at five different sampling resolution settings: 0.1 mm, 0.5 mm, 1.0 mm, 2.0 mm, and 4.0 mm, performing ten scans for each setting. The data was divided into systematic deviation analysis and measurement repeatability analysis (standard deviation) as a function of sampling density.

3.1 Analysis of Systematic Deviations and Mean Values

Analysis of Table 1 showed that the greatest deviation of the measured mean value from the nominal value was recorded at the lowest sampling density, i.e., at a resolution of 4.0 mm. In the case of Sphere 2, the measured value of 38.163 mm (nominal 38.102 mm) deviated by +0.061 mm. A similar trend, albeit on a smaller scale, was observed for Sphere 1 (+0.028 mm) and the distance between the centers of the spheres (+0.024 mm). As the sampling density increased (resolution decreased to 1.0 mm), the average value measured for all features rapidly approached the nominal value. For a resolution of 1.0 mm, the average values measured for local and global features remained in a very narrow band close to zero, showing minimal systematic deviations. Additionally, the

Table 1. Mean values of the examined dimensions at different sampling resolutions.

Mean values, mm			
Sampling resolution, mm	Distance between sphere centers (Nominal = 644.563 mm)	Sphere 1 (Nominal = 38.106 mm)	Sphere 2 (Nominal = 38.102 mm)
4.0	644.540	38.134	38.163
2.0	644.557	38.103	38.092
1.0	644.566	38.108	38.101
0.5	644.566	38.110	38.105
0.1	644.566	38.109	38.108

Table 2. Average Relative Deviations (Percentage)

Relative deviation, %			
Sampling resolution, mm	Distance between sphere centers (Nominal = 644.563 mm)	Sphere 1 (Nominal = 38.106 mm)	Sphere 2 (Nominal = 38.102 mm)
4.0	−0.0036	0.0735	0.1588
2.0	−0.0009	−0.0092	−0.0263
1.0	0.0004	0.0039	−0.0033
0.5	0.0005	0.0092	0.0072
0.1	0.0005	0.0066	0.0171

analysis of relative deviations (Table 2) confirms that the impact of resolution is most significant for smaller geometric features, where at the lowest sampling rate the error for Sphere 2 reached 0.1588%, while for the distance between the centres of the spheres it was only −0.0036%. The use of a resolution of 1.0 mm or finer allows the relative error for all tested features to be reduced to below 0.02%, which demonstrates the high metrological stability of the system in this respect.

3.2 The Effect of Sampling Resolution on Repeatability

Table 3. Statistical parameters for Sphere 1.

Sphere 1			
Sampling resolution, mm	Mean deviation, mm	Standard deviation, mm	Range, mm
4.0	−0.028	0.054	0.140
2.0	0.004	0.009	0.026
1.0	−0.001	0.009	0.027
0.5	−0.003	0.002	0.007
0.1	−0.002	0.004	0.010

Table 4. Statistical parameters for Sphere 2.

Sphere 2			
Sampling resolution, mm	Mean deviation, mm	Standard deviation, mm	Range, mm
4.0	−0.060	0.089	0.238
2.0	0.010	0.008	0.024
1.0	0.001	0.007	0.023
0.5	−0.003	0.006	0.020
0.1	−0.006	0.011	0.030

Table 5. Statistical parameters for distance between sphere centers.

Distance between sphere centers			
Sampling resolution, mm	Mean Deviation, mm	Standard Deviation, mm	Range, mm
4.0	0.024	0.026	0.081
2.0	0.023	0.009	0.034
1.0	0.017	0.008	0.020
0.5	0.017	0.003	0.008
0.1	0.016	0.006	0.017

A sharp increase in measurement repeatability (decrease in standard deviation) was observed when switching from 4.0 mm to 2.0 mm resolution. For Sphere 1, the standard deviation decreased from 0.054 mm to 0.009 mm (Table 3), which visually corresponds to the immediate concentration of measurements around the zero line in Fig. 3. The largest error spread (0.238 mm range) was recorded for Sphere 2 at a resolution of 4.0 mm (Table 4), which is clearly visible in Fig. 4 as a chaotic curve. Measurements of the global feature (Fig. 5) showed greater error stability. For a resolution of 1.0 mm, repeatability reached a stable level. The lowest standard deviation value for local features was achieved at a resolution of 0.5 mm (0.002 mm for Sphere 1). Further increasing the density to 0.1 mm did not improve the results, but caused a slight deterioration in repeatability (for Sphere 1, the standard deviation increased to 0.004 mm). Measurements in this range are concentrated in a narrow error band (see Figs. 3, 4, and 5, Table 5).

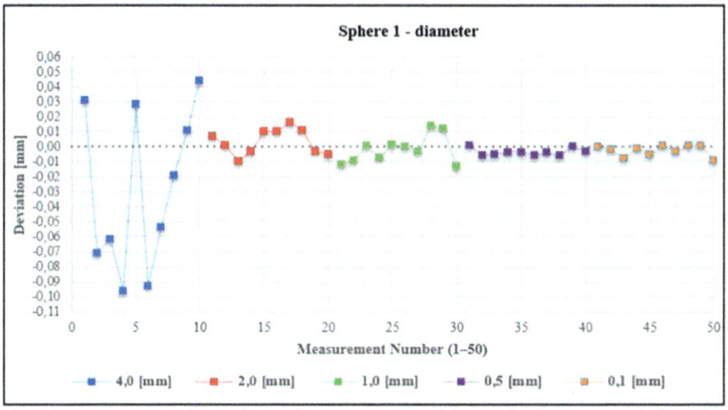

Fig. 3. Systematic deviation (mean error) of the measured sphere 1 diameter as a function of sampling resolution.

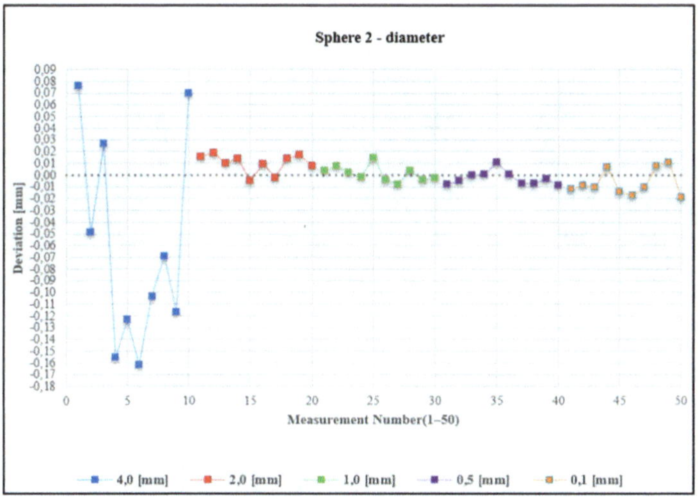

Fig. 4. Systematic deviation (mean error) of the measured sphere 2 diameter as a function of sampling resolution.

3.3 Characteristics of the Visual Representation of the STL Mesh

Observation of STL meshes (Fig. 6) shows that low sampling density (Fig. 6a) is characterized by a sparse mesh with large polygon areas. This is a structure in which the edges of the triangles are clearly visible and the shape of the sphere is visually simplified. As the resolution decreases (density increases) to 1.0 mm and 0.5 mm (Fig. 6c, d), the meshes are characterized by a significant increase in polygon density, achieving high geometric integrity. The mesh obtained at a resolution of 0.1 mm (Fig. 6e) visually presents the largest number of polygons, but surface noise texture is visible on its surface.

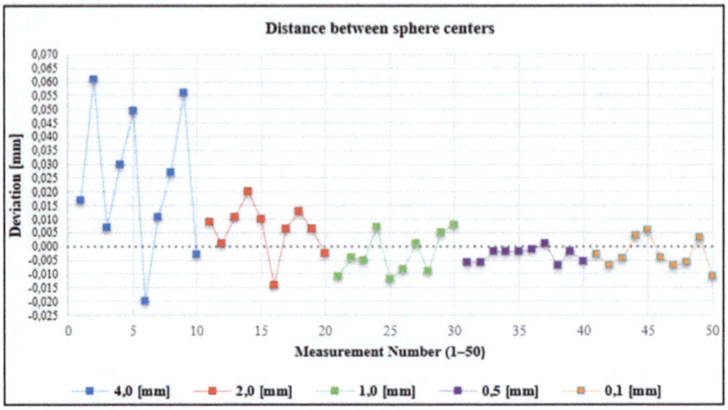

Fig. 5. Systematic deviation (mean error) of the measured distance between sphere centers as a function of sampling resolution.

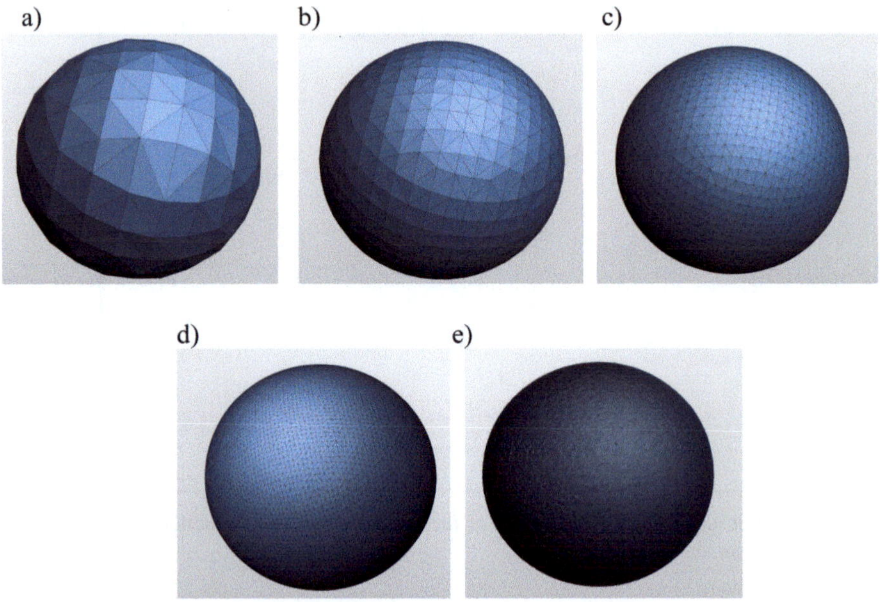

Fig. 6. STL mesh generated for sampling resolutions: a) 4.0 mm, b) 2.0 mm, c) 1.0 mm, d) 0.5 mm, e) 0.1 mm

3.4 The Influence of Sampling Resolution on the Ability to Map the Diameters and Depths of Small Holes – Supplementary Research

The results of measuring small-diameter holes (Table 6) showed that changing the resolution does not significantly affect the ability to reproduce the geometry. Up to a certain point, measurement is possible, but beyond that point it is not, regardless of the resolution settings. However, at resolutions greater than 0.3 mm, the measurement was accompanied by difficulties and not all repetitions were successful.

Table 6. Diameter difference relative to the nominal value

Hole diameter / resolution, mm	0.05	0.08	0.1	0.2	0.3	0.4	0.5
2.944	0.026 *****	0.062 *****	0.068 *****	0.016 *****	0.023 ****	0.018 ****	0.031 ***
1.944	- x	- x	0.045 **	- x	- x	- x	- x

Table 7. Depth difference relative to the nominal value

Hole diameter / resolution, mm	0.05	0.08	0.1	0.2	0.3	0.4	0.5
10.240	0.038 *****	0.040 *****	0.039 *****	0.031 *****	0.036 *****	0.037 *****	0.029 ****
9.020	0.062 *****	0.064 *****	0.066 *****	0.056 *****	0.063 *****	0.054 *****	0.060 ****
8.031	0.066 *****	0.067 *****	0.063 ****	0.066 *****	0.073 ****	0.060 *****	0.065 ****
7.027	0.073 ****	0.083 ***	0.080 ****	0.075 ****	0.073 ****	0.07 ***	0.082 ***
6.015	0.069 ***	0.079 **	0.088 **	0.087 ***	0.072 ****	0.07 **	- x
5.018	- x	- x	- x	- x	- x	- x	- x
3.866	- x	- x	- x	- x	- x	- x	- x
2.944	- x	- x	- x	- x	- x	- x	- x
1.964	- x	- x	- x	- x	- x	- x	- x

When analyzing the impact of sampling resolution on the ability to measure hole depth (Table 7), it was shown that as the diameter decreases, the difference between the average result and the nominal value increases consistently for all tested resolutions, which indicates difficulties in measuring narrower holes. At the highest resolutions of 0.4 and 0.5 mm, difficulties in obtaining data were observed, and at a resolution of 0.5 mm, a 6.015 mm hole could no longer be measured.

4 Discussion

The results of the conducted research clearly confirm that resolution and the resulting sampling density are in addition to other influencing factors [17, 18] key parameters determining the quality and stability of optical coordinate measurement. Their impact is visible both at the level of local metrics (sphere diameters) and global metrics (distance between sphere centers), and this relationship is nonlinear. The largest systematic deviations and the weakest repeatability were observed at the lowest sampling density (resolution of 4.0 mm). Both the range of deviations and the standard deviation values were many times higher in this case than for the other settings. This phenomenon is also confirmed by the graphs of measurement value changes and the assessment of STL mesh quality. A sparse mesh, consisting of large and irregular polygons, led to the flattening of local curvatures and deformation of the nominal shape of the sphere, which resulted in significant instability of the least squares fit. The high variability of the results is therefore a consequence of the insufficient number of points forming the surface model.

As the sampling density increases, there is a rapid improvement in both measurement accuracy and repeatability. Particularly favorable results were obtained for resolutions of 1.0 mm and 0.5 mm. In this range, the standard deviations reached their lowest values, and the average deviations remained stable and close to zero. The STL meshes generated for these settings were geometrically homogeneous, and the number and distribution of elements allowed for reliable fitting of primitives even with slight deviations from the ideal shape. The results obtained indicate that it is in this sampling density range that surface reconstruction and mathematical fitting algorithms work most stably.

On the other hand, no further improvement in measurement quality was observed at the highest resolution (0.1 mm). However, a slight increase in standard deviation and greater variability in results were noted. This is a typical effect of oversampling, in which additional points do not increase accuracy, but only highlight local surface noise and detection artifacts. The STL meshes for this setting were characterized by a fine, "grainy" structure, which confirms the increased noise in the data. This means that for optical scanning systems, there is a sampling resolution limit beyond which there are no metrological benefits.

In summary, the optimal sampling resolution range for the tested system is between 1.0 mm and 0.5 mm, with a resolution of 0.5 mm minimizing measurement variance and 1.0 mm providing the best compromise between acquisition time, computational load, and reconstruction quality. The results clearly show that both too low and too high sampling density leads to degradation of measurement results, although the causes in both cases are different mechanisms: simplification of the surface model and amplification of measurement noise, respectively.

In systems operating on the principle of structured light, as in any other optical measurement system [19], resolutions depend on the sensor matrix. Measurement systems/software for structured light scanning allow selection of scanning options: "Coarse Resolution (for details greater than 2.0 mm in size)," "Standard Resolution (for details greater than 1.0 mm in size)," "Fine Resolution (for details greater than 0.5 mm in size)," and "Extra Fine Resolution (for details smaller than 0.5 mm in size)," which limits the choice of resolution to four options. The authors decided to perform measurements on a system that allowed greater flexibility in parameter adjustment. Due to the analogy in sampling in all optical systems, where the continuous domain of the function I(x,y) is converted into a set of discrete values dependent on the number and area of pixels in the matrix/detector, the observed resolution thresholds will also be applicable to other systems. The validity of these observations is further supported by the experience of staff at an accredited laboratory, conducting calibrations of optical systems since 2010.

Supplementary tests have shown that sampling resolution when measuring small holes with larger values of around 0.5 mm can cause difficulties. However, the possibility of mapping is much more closely related to the geometry itself – at small diameters, measurements are no longer possible. A clear trend was also observed in which the difference between the average results and the nominal value increases with decreasing hole diameter until a threshold is reached at which measurement is no longer possible.

In the case of small features, as well as other complex geometries, e.g., deep grooves or sharp internal edges, where part of the projection is effectively "invisible" to the camera due to a limited line of sight, the problem of so-called occlusion arises (Fig. 7) — this occurs when light reflected from the surface reaches the eye but does not enter the field of view of the measurement camera [20].

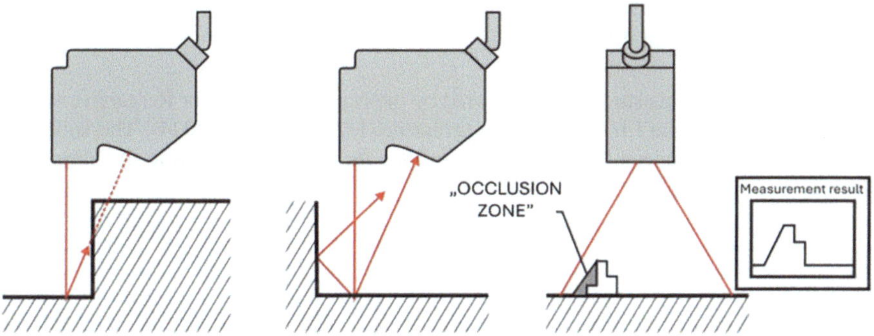

Fig. 7. Occlusion-related issues affecting measurement data: occlusion, self-occlusion, occlusion zone [20]

The phenomenon of occlusion was thoroughly analyzed in the work of Hošek [21], who pointed out significant measurement errors even with partial obstruction of the laser beam.

5 Conclusions

The conducted research clearly showed that sampling resolution is a parameter of key importance for the metrological quality of optical coordinate measurements. The obtained results confirmed that both the accuracy and repeatability of measurements are strongly dependent on the density of input data, and this influence is non-linear. The largest standard deviations and the highest measurement variability were observed at the lowest sampling density (4.0 mm resolution), which is a result of the simplification of local geometry on sparse STL meshes and the instability of the surface fitting algorithm.

It has been shown that the optimal sampling resolution range is between 1.0 mm and 0.5 mm, where the standard deviation values are minimized and the average deviations remain stable. A resolution of 0.5 mm provides the highest repeatability, while 1.0 mm is a good compromise between reconstruction quality, acquisition time, and computational load. At the same time, the existence of a saturation threshold has been confirmed: increasing the resolution to 0.1 mm does not improve the results and, in some cases, leads to their deterioration due to the registration of surface noise and reconstruction artifacts.

The presented results provide practical guidelines for users of mobile optical scanners, enabling them to make informed choices about scanning parameters depending on tolerance requirements and planned measurement time. The presented relationships also provide a basis for further research on methods for optimizing optical measurements. In particular, future research should include an analysis of the relationship between sampling density and the minimum recognizable geometric dimension, as well as the adjustment of data density to the resolution of the CAD model. The obtained results are part of a developing Expert system for estimating the uncertainty of optical measurements. The long-term goal is to develop a universal matrix for selecting scanning parameters for different tolerance classes and types of optical systems, ensuring optimization of both quality and measurement efficiency.

Acknowledgements. This study was supported by the National Center for Research and Development, Poland, under the LIDER program (grant no. LIDER15/0306/2024). The authors also acknowledge the Laboratory of Coordinate Metrology for providing access to the measurement facilities.

References

1. Wieczorowski, M., Trojanowska, J.: Towards metrology 4.0 in dimensional measurements. J. Mach. Eng. **23**(1), 100–113 (2023)
2. Stojadinovic, S.M., Majstorovic, V.D., Gąska, A., Sładek, J., Durakbasa, N.M.: Development of a coordinate measuring machine – based inspection planning system for industry 4.0. Appl. Sci. **11**, 841 (2021)
3. Taatali, A., Sadaoui, S.E., Louar, M.A., Mahiddini, B., Catalucci, S., Savio, E.: A novel study on camera calibration and deep learning based processing techniques for efficient machine vision inspection. Meas. Sci. Technol. **36**(8), 085002
4. Catalucci, S., et al.: Smart optical coordinate and surface metrology. Meas. Sci. Technol. **34**(1), 012001 (2022)
5. Guidi, G., Russo, M., Magrassi, G., Bordegoni, M.: Resolution characteritazion of 3D cameras. In: SPIE 7239, Three-Dimensional Imaging Metrology, 72390O (2009)
6. Beraldin, J.-A., Mackinnon, D., Cournoyer, L.: Metrological characterization of 3D imaging systems: progress report on standards developments. In: 17 International Congress of Metrology (2015)
7. Rachakonda, P., Cheok, G., Franaszek, M., Saidi, K.: Evaluating the depth resolution of 3D sensors for manufacturing automation applications. Measurement. **220**, 113307 (2023)
8. ISO 10360-8:2013.: Geometrical product specifications (GPS) — Acceptance and reverification tests for coordinate measuring machines (CMM) — Part 8: CMMs with optical distance sensors

9. ISO 10360-13:2021.: Geometrical product specifications (GPS) — Acceptance and reverification tests for coordinate measuring systems (CMS) — Part 13: Non-contact distance-range imaging systems
10. Liu, Y., Yu, D., Du, H., Chen, T.: Measurement of form and position error of small-diameter deep holes based on collaboration between a lateral confocal displacement sensor and helical scanning. Sensors. **25**(22), 6863 (2025)
11. Smyczyńska, L.A., Wieczorowski, M., Jakubowicz, M., Gapiński, B.: Simulation of influence of diameter and other circle parameters on results of incomplete round profile testing. Adv. Sci. Technol. Res. J. **19**(7), 151–163 (2025)
12. Poroskun, I., Rothleitner, C., Heißelmann, D.: Structure of digital metrological twins as software for uncertainty estimation. J. Sens. Sens. Syst. **11**, 75–82 (2022)
13. Wright, L., Davidson, S.: Digital twins for metrology; metrology for digital twins. Meas. Sci. Technol. **35**(5), 051001 (2024)
14. Ostrowska, K., Gąska, A., Kupiec, R., Gromczak, K., Wojakowski, P., Sładek, J.: Comparison of accuracy of virtual articulated arm coordinate measuring machine based on different metrological models. Measurement. **133**, 262–270 (2019)
15. Ostrowska, K., Gaska, A., Sładek, J.: Determining the uncertainty of measurement with the use of a virtual coordinate measuring arm. Int. J. Adv. Manuf. Technol. **71**(1–4), 529–537 (2014)
16. Confalone, G.C., Smits, J., Kinnare, T.: 3D Scanning for Advanced Manufacturing, Design, and Construction. O'Reilly Media, Sebastopol (2023)
17. Bilušić, M., Olivari, L.: Assessment of process chain suitability of the optical 3D measuring system by using influencing factors for measurement uncertainty. J. Ind. Inf. Integr. **41**, 100654 (2024)
18. Lagudi, A., Severino, U., Barbieri, L., Bruno, F.: A practical methodology for accuracy and quality evaluation of structured light systems in automotive inspection. Machines. **13**(7), 576 (2025)
19. Shan, S., Zhao, F., Li, Z., Luo, L., Li, X.: A comprehensive review of optical metrology and perception technologies. Sensors. **25**(22), 6811 (2025)
20. Ostrowska, K., Owczarek, D., Jędrzejak, I., Felho, C.: Comparative analysis of the capabilities of reverse engineering systems in reconstruction the geometry of machined components. In: Development in Machining Technology: Scientific-Research Reports, pp. 33–42. Monograph Cracow University of Technology (2025)
21. Hošek, J., Linduška, P.: Simple modification of a commercial laser triangulation sensor for distance measurement of slot and bore side surfaces. Sensors. **21**(20), 6911 (2021)

Precision Alignment and Geometrical Verification of Tire Building Machines Using Laser Tracker Systems

Marcin Moskwa[1,2](✉), Bartosz Gapiński[1], and Michał Jakubowicz[1]

[1] Faculty of Mechanical Engineering, Poznan University of Technology, Poznan, Poland
marcin.moskwa@doctorate.put.poznan.pl
[2] Bridgestone, Poznan, Poland

Abstract. This study investigates the geometrical integrity of a tire building machine utilizing a high precision laser tracker coordinate measuring system. In the study, the focus was placed on the front section of the machine, which is critical for the key stages of raw tire production, including materials application, transfer, and shaping. A series of measurements was carried out to assess the co-axiality, parallelism, and positional accuracy of critical components such as drums, transfer rings, and the bead feeder. The main objective was to evaluate the advantages and capabilities of the laser tracker method compared to conventional measurement techniques. The analysis was performed within a unified machine reference system, allowing precise and repeatable evaluation of large machine geometry. The empirical results demonstrate that the laser tracker facilitates autonomous verification of geometrical parameters without intermediate setups, significantly reducing measurement time and uncertainty. Furthermore, the study revealed that laser tracker measurements provide valuable insights into the deflection behavior of shaping drum, the alignment of transfer mechanisms, and the symmetry of the bead feeding system. These capabilities make the laser tracker a powerful diagnostic and calibration tool in industrial large-scale metrology. The application of this method in tire building processes improves assembly accuracy, reduces geometric errors, and enhances product uniformity.

Keywords: Laser Tracker · Tire Building Machine · Large-Scale Metrology · Machine Alignment

1 Introduction

Tire uniformity, characterized by radial (RFV) and lateral (LFV) force variations, represents a critical safety and performance requirement in the automotive industry. These variations originate primarily from structural defects incurred during the green tire building stage on the tire building machine (TBM) [1, 2]. Geometrical inaccuracies, particularly drum misalignment, are preserved throughout the curing process, directly inducing a non-uniform distribution of stiffness and mass in the cured tire [3]. Achieving TBM geometric tolerances in the order of tens of micrometers on large-scale industrial machinery necessitates a transition from traditional metrology to advanced large-volume coordinate measurement systems (CMS) [4].

M. Hryb et al. (Eds.): MANUFACTURING 2026, LNME, pp. 76–91, 2026.
https://doi.org/10.1007/978-3-032-23038-6_6

Recent studies have demonstrated the successful application of laser tracker (LT) systems in large-scale industrial metrology, particularly for the geometric verification and alignment of large assemblies. Practical applications include on-site dimensional inspections in shipbuilding and heavy industry, where LT systems are employed to establish global reference frames and to compare different large-scale metrology systems [5], as well as the alignment and calibration of industrial robots and large machining setups across extended workspaces [6, 7]. These studies confirm that laser trackers provide high spatial accuracy and flexibility for large-volume measurements.

However, laser tracker-based measurements are inherently constrained by line-of-sight requirements and feature accessibility [8]. Components that are not visible from a single tracker position cannot be measured directly, necessitating tracker relocation and the use of multi-station measurement strategies with coordinate transformations [4]. The resulting measurement uncertainty is significantly influenced by station repositioning accuracy, target redundancy, and network geometry [9]. These limitations are particularly relevant for complex manufacturing machines with enclosed structures and rotating subsystems, such as tire building machines. In this context, the present paper investigates laser tracker-based geometric verification of a TBM. The proposed approach utilizes a machine-specific reference system to evaluate the co-axiality, parallelism, and positional deviations of functional components, while explicitly accounting for the visibility and accessibility constraints inherent to such machines.

1.1 Critical Analysis of Conventional Metrological Methods and Tools in Geometric Inspection of Large-Scale Industrial Systems

1.1.1 Evaluation of Conventional Metrological Instruments

The geometric alignment of critical machine components requires advanced metrological procedures. Historically, the industry has relied on conventional tools; however, these instruments demonstrate fundamental limitations as manufacturing technologies evolve toward increasingly stringent geometric tolerances and higher precision requirements [10]. Common conventional tools employed in the machinery industry include calipers, micrometers, dial indicators, and surveying equipment such as optical or electronic theodolites [11]. These instruments are inherently limited to 1D or 2D measurements. Consequently, a comprehensive spatial reconstruction of the Tire Building Machine (TBM) geometry, incorporating features such as shaft co-axiality or axis parallelism, necessitates multiple, sequential measurement setups and subsequent numerical triangulation. This iterative process is highly susceptible to the accumulation of transfer errors [9]. Furthermore, the measurement uncertainty associated with these methods frequently exceeds the tolerances mandated by Geometrical Product Specifications (GPS) standards [12, 13].

1.1.2 Comparative Analysis: Laser Tracker Versus Conventional Metrological Methods

The transition to techniques based on coordinate measuring systems, such as the laser tracker, is necessary. The LT system effectively mitigates the inherent limitations of traditional measurement techniques, as synthesized in Table 1.

Table 1. Critical comparison of metrology methods (LT vs. conventional) in large scale system control [8, 14–16].

Metrological feature	Conventional tools (caliper, theodolite, indicator)	Laser tracker
Measurement dimension	Typically limited to 1D or 2D. Full 3D reconstruction requires multiple setups and numerical triangulation, increasing transfer error risk.	Full 3D measurement within a unified coordinate measuring system from a single setup.
Measurement uncertainty	Higher uncertainty; typically ranges from ±0.1 mm to ±1 mm per meter for large components. Inadequate for verifying tolerances ≤0.5 mm over long distances.	Significantly lower uncertainty; typically around ±0.05 mm in large-volume applications.
Environmental compensation	Usually absent or manually corrected (e.g. for thermal expansion).	Active compensation for temperature, air pressure, and humidity, automatically applied during measurement.
Automation and efficiency	Manual operation and data recording; requires repositioning and re-measurement; long inspection time.	Automated data acquisition, point registration, and direct comparison with CAD; significant time reduction.

The adoption of CMS technology is a fundamental prerequisite for achieving optimal tire uniformity in large-scale industrial systems. Furthermore, the inherent mobility of the laser tracker facilitates high-precision verification directly at the machine's installation site, supporting the geometric alignment of large assemblies within their operational environment. This methodology eliminates the necessity for component relocation and ensures metrological consistency under actual working conditions.

2 Research Problem and Methodology

2.1 Object of Study Description: the Tire Building Machine

The object of study is a three-drum tire building machine, the construction of which is fundamental to analyzing geometric errors that negatively impact tire uniformity [17]. A conceptual overview of the machine's configuration is presented in Fig. 1 [18]. The green tire assembly process, which refers to the tire before curing, comprises several sequential and concurrent stages, depending on the specific machine configuration and operational capacity. In industrial practice, the TBM represents one of the most complex and geometrically sensitive assemblies within the tire production line. Its mechanical performance is a primary determinant of the dimensional stability, uniformity, and dynamic balance of the final product. Consequently, a rigorous understanding of its structural architecture and the spatial relationships between its primary functional components is essential for both process optimization and advanced maintenance diagnostics.

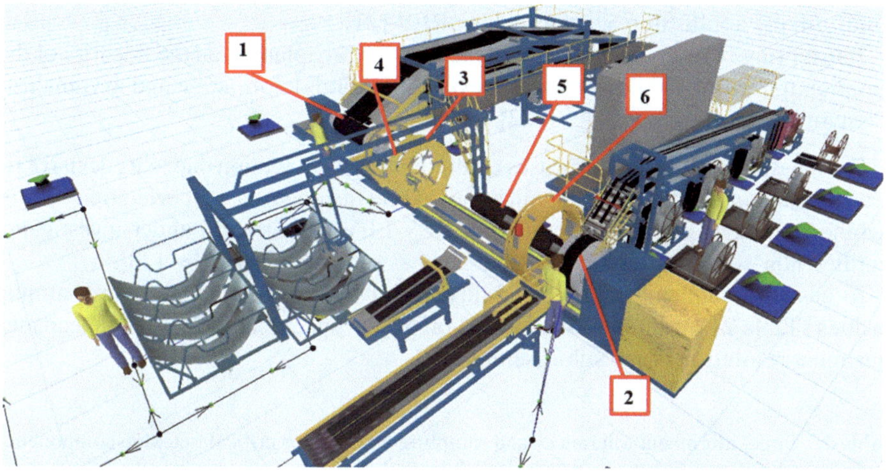

Fig. 1. Three drum tire building machine – conceptual view: 1 - carcass drum; 2 - belt and tread (B&T) drum; 3 - carcass transfer ring; 4 - bead feeder; 5 - shaping drum; 6 - B&T transfer ring [18]

The green tire construction sequence involves the following stages [17, 19]:

- Carcass package construction (left): application of the innerliner, sidewalls, and body ply onto the carcass drum (1).
- Belt-and-tread (B&T) package construction (right): Application of belt plies, cap ply, and tread onto the B&T drum (2).
- Transfer and joining: the carcass transfer ring (3) moves the carcass for bead (4) application and then to the shaping drum (5). Concurrently, the B&T transfer ring (6) positions the tread package over the shaping drum (5) for final joining and shaping.
- Final stage: the green tire is removed from the shaping drum (5) and transported to the curing area.

2.1.1 Geometric Requirements and Measurement Strategy

The successful building of the green tire requires precision in the positioning and transfer of components Since all functional elements traverse along a common machine bed, this structure dictates the operational positions of individual components. Given the distributed and multi-stage nature of the TBM design, the geometric integrity of several key subsystems must be ensured to maintain consistent tire quality. Any deviation in mutual alignment can accumulate throughout the production cycle, resulting in material placement errors, splice irregularities, or asymmetrical shaping. Key functional assemblies requiring rigorous geometric control include:

- The machine bed (main frame): This structure serves as the primary geometric datum for the entire system; any deviations in its flatness or straightness propagate as positioning errors across all moving elements [20].
- Building drum axes (1, 2, 5): Precise axial alignment and the minimization of radial and axial runout are critical, as these errors constitute the primary cause of non-uniform ply application and subsequent RFV [21].
- Transfer rings (3): The parallelism of the transfer ring planes and the linearity of their motion trajectories are essential to prevent structural deformation and asymmetrical assembly of material packages [22].

The measurement strategy focuses on the TBM's front section due to its decisive role in carcass shaping and material application. While the quantitative correlation between machine deviations and final tire quality (RFV/LFV) is currently under investigation, identified non-conformities are strictly treated as process instabilities [23].

To ensure manufacturing predictability, target alignment tolerances and sampling densities (Table 2) were established based on design specifications and the mechanical adjustment resolution of the sub-assemblies.

Table 2. Target alignment tolerances and sampling strategy for critical machine components.

Feature	Machine bed	Building drums (1, 2, 5)	Bead feeder system	Transfer ring segments
Measured parameter	Straightness and flatness	Co-axiality and runout	Co-axiality and symmetry	Radial distance and parallelism
Target alignment tolerance [mm]	–	0.25	0.5/0.25	0.5
Minimum sampling points	20	15	6	2 per each

The sampling density was selected to ensure the mathematical stability of the geometric feature fitting. For the machine bed and building drums, a higher point density was utilized to minimize the influence of local surface irregularities through a least-squares approximation of the functional axes.

2.2 Technical Specification and Metrological Properties of the Laser Tracker System

This study utilized the FARO Vantage laser tracker system designed for large-scale metrology. The selection of this instrument was necessitated by the requirement for high-precision measurements (in the range of hundredths of a millimeter) within a working volume typical of TBMs (up to several meters) [8].

The FARO Vantage system determines 3D coordinates by tracking a spherically mounted reflector (SMR) using angular encoders and Absolute distance meter (iADM/TruADM) technology [24, 25]. The selection of the LT is justified by its MPE

(20 μm + 5 μm/m) which significantly exceeds the requirements (a target of 0.25 mm). This provides a substantial safety margin for measurement accuracy at the micrometer level [26]. Such precision is maintained across the system's operational volume, extending up to several dozen meters, which is critical for the metrological verification of TBM assemblies [27].

Beyond its intrinsic precision, a strategic advantage of the laser tracker is its capacity for measurement standardization across various production facilities. By digitally recording the 3D coordinates of all critical TBM subsystems, the LT system facilitates a digitized comparison of the geometric status of different machines within a plant or a global network [5]. Unlike analog measurements, which yield incomparable partial results, LT data is integrated into a global coordinate system and can be directly compared against a CAD model or golden standard. This enables the rapid identification of sub-assemblies requiring maintenance and minimizes process variability between individual workstations [28]. Achieving a repeatable and uniform geometry across the entire machine fleet is a fundamental prerequisite for the stable production of high-uniformity tires [23].

2.3 Machine Reference System Definition

Precise alignment and geometric inspection of the TBM require anchoring all measurements to a stable and repeatable three-dimensional coordinate system. Given that critical geometric tolerances are referenced to internal factory standards rather than the nominal CAD model, a dedicated Machine Reference System (MRS) was established. This system is defined by the machine's fundamental functional features, which dictate its primary operational axes (see Fig. 2) [29].

The MRS was defined from the machine's functional features and subsequently aligned to the laser tracker using metrology software through the following sequential datum definitions:

- Origin definition: The coordinate system origin was determined by measuring the cylindrical geometric feature (the bearing flange) to compute its geometric center. The location of the center of this bearing flange (on the right side of the machine, near the B&T drum) was adopted as the zero-point for the MRS.
- Vertical datum definition (Y-axis): The Y-axis (vertical axis) was defined based on a gravity-referenced leveling plane determined using the tracker's built-in electronic levelling function. This plane, which is perpendicular to the vector of gravity and aligned (translated) to the established bearing center, became the primary datum for height measurements and vertical deviations. The Y-axis is perpendicular to this horizontal plane.
- Defining the main machine axis (Z-axis): The Z-axis, extending along the machine bed, was constructed by measuring multiple surface points on the bed structure. A regression line was fitted to these points and projected onto the leveled horizontal plane. This line was then geometrically translated to pass through the MRS origin.
- Defining the X-axis: The X-axis (the transverse axis) was automatically generated as perpendicular to the Y-axis and the Z-axis, ensuring a right-handed orthogonal coordinate system

This established MRS ensures that all subsequent measurements, including the co-axiality and parallelism of mobile components, are precisely referenced to the functional geometric axis of the machine. The ability to switch between Cartesian and cylindrical coordinate views in the LT software is crucial for the measurement and correction of axial co-axiality, plane parallelism, and radial distances, thus facilitating diagnostics.

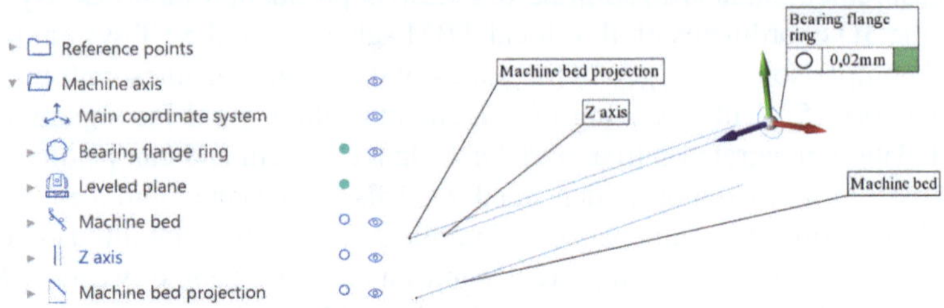

Fig. 2. Machine Reference System based on features.

2.4 Measurement Uncertainty and Repeatability

To ensure the reliability of the geometrical verification, a combined uncertainty budget was estimated. The total measurement uncertainty Uc was calculated using the root sum square method, considering the primary components presented in Table 3: instrument MPE, SMR centering tolerances, feature fitting residuals (least-squares method), and station relocation (drift) errors:

$$U_c = \sqrt{u_{instr}^2 + u_{SMR}^2 + u_{fit}^2 + u_{reloc}^2} \quad (1)$$

Based on industrial conditions and a working range of 1.5–5.0 m, the estimated combined uncertainty was Uc ~ 70 µm. This value satisfies the metrological requirement for verifying the machine's tightest tolerances (0.25 mm), maintaining a precision-to-tolerance ratio of approximately 1:3,5. This confirms the laser tracker's capability to detect geometric deviations that remain undetectable by conventional alignment tools. To minimize thermal influence, measurements were conducted under stable ambient conditions, and the duration of each cycle was kept minimal to neglect machine thermal drift. Repeatability was verified through three independent measurement cycles of the main drum's axis, yielding a standard deviation of s < 25 µm, which confirms the robustness of the methodology for TBM diagnostics.

Table 3. Estimated uncertainty contributions (relative geometry)

Uncertainty component	Source/basis	Value [µm]
Instrumental (u_{inst})	MPE (20 µm + 5 µm/m)	< 45
SMR Centering (u_{SMR})	Manufacturer specs	± 12
Feature Fitting (u_{fit})	Least-squares residual	20–40

(continued)

Table 3. (*continued*)

Uncertainty component	Source/basis	Value [μm]
Relocation (u_{reloc})	Network transformation	≤ 40
Combined (U_c)	RSS estimate	~70

3 Research Problem

The rigorous geometric verification of a tire building machine presents a significant metrological challenge due to the machine's substantial dimensions, the limited accessibility of critical internal features, and the complex spatial relationships between its various rotating assemblies. Conventional measurement methodologies, typically relying on mechanical gauges or manual referencing, fail to provide the requisite precision, traceability, or repeatability necessary to assess fine geometric deviations between major functional components. To address these limitations, the application of a laser tracker integrated with dedicated metrology software has been used. This methodology facilitates not only the high-precision measurement of critical geometric elements but also enables a standardized direct comparison between multiple machines of identical configuration, thereby supporting both predictive diagnostics and precise mechanical alignment.

The analysis focuses on the evaluation of geometric relationships such as co-axiality, radial runout, and parallelism between the principal structural and functional components of the TBM. The critical components targeted for measurement and potential corrective alignment include the carcass drum, shaping drum, B&T drum, carcass and B&T transfer rings, the machine bed, and the bead feeder system.

Consequently, the core research problem involves the development and validation of a feature based alignment methodology using a laser tracker to establish a stable and traceable machine reference system. Defining this reference frame represents one of the main metrological challenges, as it requires the precise determination of the machine's geometric base through leveling, axis orientation, and feature correlation. Once established, the MRS serves as the foundation for repeatable assessment and standardized adjustment of large-scale TBM assemblies.

4 Results and Discussion

In the conducted measurement studies, the front section of the tire building machine was analyzed, the area where the key stages of the green tire building process take place. In this zone, individual layers of materials are applied onto the drums, transferred over the shaping drum, and joined together. Finally, on the shaping drum, the complete green tire is shaped.

The application of a laser tracker enabled a comprehensive geometric analysis of these processes within the predefined machine reference system. A series of measurements were performed, covering the key machine components on both the carcass side and the B&T side. However, this article focuses mainly on those measurement stages that most clearly demonstrate the advantages and capabilities of the laser tracker compared with conventional measurement techniques. This approach not only documents the geometric condition of the machine but also serves as a practical demonstration of the benefits of modern large-scale metrology in industrial applications.

4.1 Measurement of the Machine Bed

The initial phase of the experimental verification involved the metrological characterization of the machine bed. Preliminary analysis determined that 20 discrete measurement points distributed along the bed surface are sufficient to establish the primary machine axis with high statistical confidence. For clarity of visualization, 10 representative points are illustrated in Fig. 3.

This measurement was essential as it enabled both the definition of the machine's main axis and the evaluation of the movement of the transfer rings relative to this axis. In the analyzed case, the carcass transfer ring, moving from the left side of the machine toward the center, initially shifts downward and then ends its motion approximately 0.2 mm above its starting position. This corresponds to the natural movement path during carcass transfer to the shaping drum. Similarly, the B&T transfer ring, moving from the right side of the machine, descends by about 0.8 mm as it approaches the shaping drum area, where the B&T package is positioned and joined with the carcass.

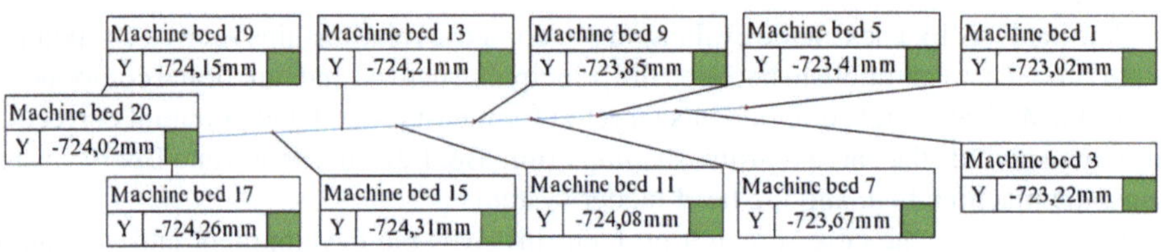

Fig. 3. Overview of the machine bed measurement.

The movement of the carcass transfer ring is not critical since the shaping drum ultimately defines the final position of the carcass. At the moment the carcass is captured by the shaping drum, the material layers are not yet joined. However, in the case of the B&T transfer ring, this movement is significant, as it affects the concentricity of the material packages, potentially leading to misalignment between the carcass and the B&T package.

4.2 Verification of Drum Co-axiality

Co-axiality of the rotating assemblies is a fundamental determinant of tire building quality. Under conventional maintenance protocols, drum axes are rarely verified directly, as dial gauge measurements are typically relative to an unverified reference drum. The application of a laser tracker enables the direct verification of drum co-axiality within the unified MRS. The results, presented as the deviation of geometric centers relative to the machine axis in the X and Y planes, are shown in Fig. 4

A measured deviation of 0.79 mm in the carcass drum position was identified This magnitude of error significantly impacts all subsequent measurements that utilize the drum as a reference datum. When using a laser tracker, measurements and subsequent adjustments are independent of the drum alignment, increasing both reliability and repeatability. The identified 0.79 mm deviation confirms that the laser tracker effectively detects non-conformities exceeding plant thresholds (0.25 mm), providing precise data for necessary mechanical corrections.

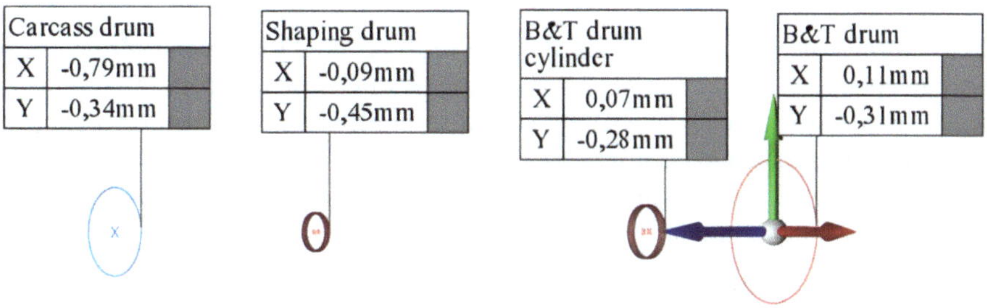

Fig. 4. Drum alignment results.

4.3 Comparative Analysis of the B&T Transfer Ring Geometry

The B&T transfer ring comprised 11 segments and plays a crucial role in the green tire building process, operating in two primary working positions:

- capturing the B&T package over the B&T drum, and
- bonding the carcass with the B&T package over the shaping drum.

Radial measurements performed using the laser tracker were compared with those obtained via a conventional dial gauge (see Table 4).

Table 4. Comparative radial distance measurements for the B&T transfer ring.

	Shaping drum position		B&T drum position	
Segment number	Laser tracker [mm]	Dial gauge [mm]	Laser tracker [mm]	Dial gauge [mm]
1	0,17	−0,01	−0,04	-0,28
2	0,25	-0,57	-0,32	-0,93
3	0,3	−1,07	−0,66	−1,52
4	0,44	−1,29	−0,81	−1,82
5	0,41	−1,43	−0,95	−2,01
6	0,25	−1,43	−0,97	−1,9
7	−0,17	−1,37	−1,09	−1,73
8	0,08	−0,44	−0,43	−0,65

(continued)

Table 4. (*continued*)

	Shaping drum position		B&T drum position	
Segment number	Laser tracker [mm]	Dial gauge [mm]	Laser tracker [mm]	Dial gauge [mm]
9	0,01	−0,06	−0,16	−0,11
10	−0,1	−0,08	−0,09	−0,15

In traditional measurements, the dial gauge must be fixed to reference drums, requiring two independent measurements for both ring positions. In contrast, the laser tracker allows both positions to be measured within a single setup, using previously measured machine bed data and coordinate transformation to the respective drum positions. This approach significantly reduces measurement time and eliminates errors caused by repositioning or misalignment of the dial gauge.

The comparative analysis focused on the radial distance deviations across the 11 segments of the transfer ring. The dial gauge measurements were inherently referenced to the first segment, whereas the laser tracker data were established relative to the defined machine axis (MRS), with the first segment utilized as a baseline for normalization. Consequently, negative values presented in Table 4 denote a reduction in radial distance relative to the first (reference) segment. The most significant deviations were identified in segments positioned below the machine's functional axis, specifically segments 5 through 7. In these instances, the discrepancies between the two metrological methods exceeded 1.5 mm.

4.4 Additional Analysis - Influence of Shaping Drum Deflection

The shaping drum axis was evaluated relative to the machine's primary axis at both the proximal and distal ends of the drum mounting (Fig. 4). Since the tire shaping process occurs in the bead application zone, an additional experiment was performed using a reference bead wire. A measurement point was defined on the drum in the upper vertical position; after a 180° rotation, the same point was measured in the lower vertical position. This procedure was repeated for two distinct points (Fig. 5).

The results showed radial distance increases of 0.96 mm and 1.67 mm for points 1 and 2, respectively, indicating significant shaft deflection in the central region of the drum. These measurements verified that shaft deflection potentially compromises the drum's operational geometry, which may lead to off-center bead wire positioning. Consequently, transfer ring measurements performed with a dial gauge mounted directly on the drum may yield inaccurate results relative to the machine axis.

Machine calibration therefore depends on drum type and its susceptibility to deflection. In contrast, laser tracker measurements are independent of drum mass and geometry, facilitating consistent and objective comparisons between different tire building machines.

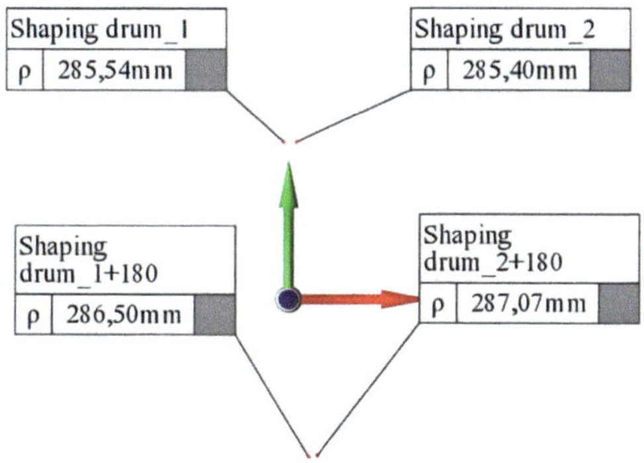

Fig. 5. Deflection measurement of the shaping drum axis in the upper and lower working positions.

4.5 Measurement and Analysis of the Bead Feeder

The bead feeder system was evaluated to verify the co-axiality of the bead application process relative to the carcass. A reference bead was measured at two critical stages: the application position and the pickup position by the carcass transfer ring. The analysis involved determining the geometric centers of circles reconstructed from measurement points on both the left (L) and right (R) sides of the feeder. The results are summarized in Table 5.

Table 5. Geometric centers of the reference bead during different application stages

	Bead feeder L	Bead CCS application L	Bead feeder R	Bead CCS application R
X [mm]	1,10	0,77	0,92	0,87
Y [mm]	−0,45	−0,37	0,24	0,14

The measurements identified positional discrepancies in the X-axis (0.05–0.33 mm) and Y-axis (0.08–0.10 mm). While these values do not significantly impair pickup accuracy, the results confirm that precise feeder alignment is indispensable for maintaining concentricity during bead insertion.

The secondary phase of the analysis focused on the positional symmetry between the left and right feeder assemblies. The feeder's application mechanism is mechanically coupled, meaning that adjustment on one side should automatically ensure correct positioning on the opposite side. Figure 6 presents two configurations of the bead: before and after adjustment.

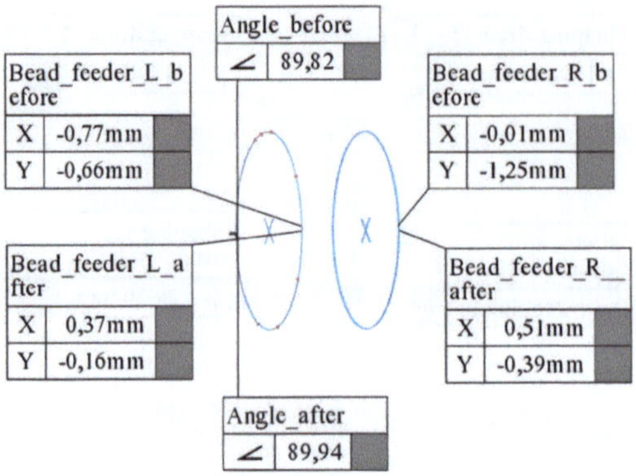

Fig. 6. Bead feeder system before and after adjustment.

Before adjustment, the measured bead positions differed by 0.76 mm in the X-axis and 0.59 mm in the Y-axis between the right and left sides. The feeder position was evaluated by constructing a circle from the measurement points of the reference bead surface. Using the laser tracker software, a reference plane was then created from the measured points, and the angle between this plane and the machine's main axis was calculated. Ideally, this angle should equal 90°. In the analyzed case, it was 89.82° before adjustment and improved to 89.94° after calibration. This correction reduced positional differences between the left and right bead application sides, as summarized in Table 6.

Table 6. Bead feeder alignment results.

	Bead feeder L before	Bead feeder R before	Bead feeder L after	Bead feeder R after
X	−0,77	−0,01	0,37	0,51
Y	−0,66	−1,25	−0,16	−0,39

5 Conclusions

The conducted research confirms the high efficacy of the laser tracker system for the geometric diagnostics and alignment verification of tire building machines. The results clearly demonstrated the superiority of this technology over conventional measurement methods in terms of accuracy, repeatability, and the capacity for comprehensive geometric analysis within a unified machine reference system. The application of the laser tracker facilitated:

- Direct verification of co-axiality, parallelism, and runout of key working components without intermediate measurements.
- Analysis of transfer ring movements and their influence on the concentricity of carcass and B&T bonding.

- Identification of shaping drum shaft deflection and its effect on the accuracy of traditional measurements.
- Symmetry evaluation of the bead feeder system, including the precise calibration of angular and linear deviations.

The indicate that traditional dial gauge methods are inadequate for modern precision requirements, particularly when tolerances are in the range of hundredths of a millimeter or over extended measurement volumes. In contrast, the laser tracker provides high spatial precision and repeatability, enabling the simultaneous analysis of multiple geometric parameters within a single metrological setup. Furthermore, the integration of dedicated metrology software facilitates automated coordinate transformations, which significantly reduces inspection cycles and eliminates errors associated with manual setups. From an industrial perspective, a decisive advantage is the capacity to digitally compare various machines within a unified reference system, thereby establishing a foundation for the comprehensive metrological standardization of the production environment.

In conclusion, the laser tracker measurement method is an effective and robust tool for controlling and correcting the geometry of large industrial machines. Its application in tire building processes enhances assembly accuracy, reduces geometric errors, and improves the uniformity of the final product. Furthermore, the developed measurement protocols and the unified reference system are scale independent and transferable to various tire building configurations. This includes second-stage machines or alternative building sequences, where maintaining global axis integrity is equally vital.

Acknowledgements. The authors thank the Polish Ministry of Education and Science for financial support (Applied Doctorate Program, No. DWD/ 6/0175/2022). This research was funded by the Ministry of Science and Higher Education grant no. 0614/SBAD/1603

References

1. Du, M., Sun, P., Zhou, S., Huang, H., Zhu, J.: A study on the influence of tire speed and pressure on measurement parameters obtained from high-speed Tire uniformity testing. Vehicles. **2**(3), 559–573 (2020). https://doi.org/10.3390/vehicles2030031
2. Starrett-Bytewise.: Five Ways to Improve Tire Uniformity at the Tire Building Machine, Laser measurement solutions. Accessed: Nov. 13, 2025. [Online]. Available: https://www.starrett.com/docs/default-source/bytewise/tire-industry-docs/3003%2D%2D-5-ways-to-improve-tire-uniformity.pdf?sfvrsn=9690bd7e_7
3. Wang, Y., Li, Z., Li, Z., Wang, Y.: Finite element simulation of radial tire building and shaping processes using an Elasto-Viscoplastic model. Comput. Model. Eng. Sci. **135**(2), 1187–1207 (2023). https://doi.org/10.32604/cmes.2022.022596
4. Peggs, G.N., et al.: Recent developments in large-scale dimensional metrology. Proc. Inst. Mech. Eng. B J. Eng. Manuf. **223**(6), 571–595 (2009). https://doi.org/10.1243/09544054JEM1284
5. Maisano, D.A., et al.: Dimensional measurements in the shipbuilding industry: on-site comparison of a state-of-the-art laser tracker, total station and laser scanner. Prod. Eng. **17**(3–4), 625–642 (2023). https://doi.org/10.1007/s11740-022-01170-7
6. Pantscharowitsch, M., Moser, L., Kromoser, B.: A study of the accuracy of industrial robots and laser-tracking for timber machining across the workspace. Wood Mater. Sci. Eng. **20**(1), 75–93 (2025). https://doi.org/10.1080/17480272.2024.2324437

7. Zheng, Y., Liu, W., Zhang, Y., Ding, H., Li, J., Lu, Y.: Laser in-situ measurement in robotic machining of large-area complex parts. Measurement. **241** (2025). https://doi.org/10.1016/j.measurement.2024.115718
8. Muralikrishnan, B., Phillips, S., Sawyer, D.: Laser Trackers for Large-Scale Dimensional Metrology: a Review. Elsevier Inc. (2016). https://doi.org/10.1016/j.precisioneng.2015.12.001
9. Muelaner, J., Maropoulos, P.G.: Large scale metrology in aerospace assembly, 5th international conference on digital enterprise technology. Nantes. **773–792** (2008)
10. Schmitt, R.H., et al.: Advances in large-scale metrology – review and future trends. CIRP Ann. **65**(2), 643–665 (2016). https://doi.org/10.1016/J.CIRP.2016.05.002
11. Raghavendra, N.V., Krishnamurthy, L.: Engineering Metrology and Measurements. Oxford University Press, New Delhi (2013)
12. Geometrical product specifications (GPS) - Geometrical tolerancing - Tolerances of form, orientation, location and run-out. ISO 1101 (2017)
13. Geometrical product specifications (GPS) - Acceptance and reverification tests for coordinate measuring systems (CMS) - Part 10: Laser trackers. ISO 10360-10 (2021)
14. Zhu, X., Zheng, L., Tang, X.: Configuration optimization of laser tracker stations for large-scale components in non-uniform temperature field using Monte-Carlo method. Proc. Cirp. **56**, 261–266 (2016)
15. Flack, D., Hannaford, J.: Good Practice Guide No. 80 Fundamental Good Practice in Dimensional Metrology, National Physical Laboratory (2005)
16. Caskey, G. W. Jr. Fronczek, C. J., Phillips, J. S. D.: Laser trackers: traceability, uncertainty, and standardization a report to the CMSC, U.S. Department of commerece (1997)
17. Rockwell Automation, Tire Building Machine Reduce Design Time and Improve TCO for Tire Manufacturers, Tire – WP002A-EN-P. 2013. Accessed: Nov. 03, 2025. [Online]. Available: https://www.scribd.com/document/421403901/tire-wp002-en-p-pdf
18. Cao, M., Wang, G., Sun, L.: Research on key technology of new concept tyre building production line. Int. J. Intell. Syst. Appl. **4**(8), 46 (2012). https://doi.org/10.5815/IJISA.2012.08.06
19. Suffeleers, J.: Tyre Non-uniformities and their Effect on Chassis Vibrations. Eindhoven University of Technology (2010)
20. Machine bed support key to accuracy, Cutting Tool Engineering, 2012. Accessed: Nov. 03, 2025. [Online]. Available: https://www.ctemag.com/articles/machine-bed-support-key-accuracy
21. Mawby, W.D., O'Brien, G.P., Persyn, M.E., Traylor, J.M.: Tire manufacturing method for improving the uniformity of a tire, US Patent 6856929 B1 (2005)
22. Transfer Rings/O Rings. Accessed: Nov. 03, 2025. [Online]. Available: https://wyko.com/products/transfer-rings-o-rings
23. Moskwa, M., Gapiński, B., Jakubowicz, M.: Radial force variation parameter as a factor of tire uniformity. Bull. Polish Acad. Sci. Tech. Sci. **73**(2) (2025). https://doi.org/10.24425/bpasts.2024.152609
24. Yang, L., et al.: Automatic guidance method for laser tracker based on rotary-laser scanning angle measurement. Sensors. **20**(15), 1–17 (2020). https://doi.org/10.3390/s20154168
25. Surf's Up for Laser Trackers: How the Precision Measurement Tool Helps Wave Worshippers Catch the Ultimate Barrel, https://www.faro.com/en/Resource-Library/Article/Laser-Trackers-Helps-Catching-Waves
26. Faro Vantage Laser Trackers - Tech. Sheet, Faro Knowledge Base. Accessed: Nov. 13, 2025. [Online]. Available: https://media.faro.com/-/media/Project/FARO/FARO/FARO/Resources/2_TECH-SHEET/FARO-Vantage-Laser-Trackers/TechSheet_Vantage_ENG.pdf?rev=-1

27. Machinery for the Tire & Rubber Industry, Inter European Machinery & Engineering. Accessed: Nov. 13, 2025. [Online]. Available: https://www.intereuropean.it/files/Tire-Machinery-Catalog-2018snu.pdf
28. Han, R., Dunker, T., Trostmann, E., Xu, Z.: Calibration methods for large-scale and high-precision globalization of local point cloud data based on iGPS. Sensors. **24**(18) (2024). https://doi.org/10.3390/s24186114
29. Subirón, F.R., Castellano, P.R., Bellido, G.M.B., Nácher, S.B.: Feature-based framework for inspection process planning. Materials. **11**(9) (2018). https://doi.org/10.3390/ma11091504

Application of the LSCI Method for Assessing the Roundness Deviation of Circular Profile Segments with Different Types of Form Deviation

Lidia Smyczyńska(✉), Bartosz Gapiński, and Michał Jakubowicz

Faculty of Mechanical Engineering, Poznan University of Technology, Poznan, Poland
lidia.smyczynska@doctorate.put.poznan.pl

Abstract. One of the key challenges in geometrical quantity metrology is the evaluation of roundness deviation of rotational elements when access to the full profile is limited. In industrial practice, roundness measurements are often limited to partial contours, which can influence the resulting evaluation of roundness deviation. This study investigates the influence of the analysed arc length and its position relative to characteristic features of the profile on the value of roundness deviation determined using the Least Squares Circle (LSCI) method. Experimental investigations were carried out using a form tester on internal circular profiles with three different types of geometric deformation: oval, triangular, and quadrangular. Arc segments ranging from the full circumference of 360° down to 15° were analysed at three characteristic positions with respect to the profile geometry: at the peak, in the valley, and in an intermediate location. The results demonstrate that shortening the analysed arc length leads to a reduction in the evaluated roundness deviation for all investigated profiles. However, the magnitude and character of this effect strongly depend on the geometric form of the profile. With an increasing number of form waves, the decrease in roundness deviation becomes more gradual and the sensitivity to the position of the analysed segment is reduced. The quadrangular profile exhibited the highest stability of results and the smallest differences between measurement positions, whereas the oval profile showed pronounced variability and strong dependence on the segment location. The findings provide experimental confirmation of the authors' previous simulation-based studies and demonstrate that the LSCI method can be effectively applied to the evaluation of roundness deviation under conditions of limited profile coverage, provided that the analysed arc length is appropriately selected.

Keywords: Arc Length · Form Deviation · Incomplete Round Profile · Internal Profile · LSCI Method · Roundness · Sampling Strategy

1 Introduction

The requirements for the accuracy and efficiency of roundness measurements are primarily determined by design tolerances defined to ensure the functional performance of parts. These demands arise not only from elevated quality expectations for engineering

M. Hryb et al. (Eds.): MANUFACTURING 2026, LNME, pp. 92–105, 2026.
https://doi.org/10.1007/978-3-032-23038-6_7

components but also from the increasing complexity of their geometry, material properties, and design tolerances, which require the use of measurement strategies appropriate for the intended inspection task in both laboratory and production environments [1]. Even though real parts inevitably exhibit some deviation from the ideal circular form, it is the exceedance of specified tolerance limits that may lead to operational problems, increased friction, wear, and a deterioration of functional performance. Therefore, the evaluation of roundness deviation is of key importance for ensuring product quality and reliability [2]. In recent decades, the demands placed on coordinate metrology techniques have risen substantially due to the growing geometric complexity of components. It is estimated that approximately 70% of engineering parts possess an axis of rotational symmetry, making the control of deviations from the ideal round shape one of the primary quality criteria in precision manufacturing [3].

A review of the available methods for roundness assessment has been presented by Shen et al. [4], Singh and Arunachalam [5], as well as Moroni and Petro [6], who highlighted various approaches to data acquisition and analysis. Weckenmann et al. [1] emphasized the need for a consistent sampling strategy, as the number and distribution of measurement points strongly affect the reliability of the results. In practice, roundness can be measured using both general-purpose and dedicated form-measuring instruments [7], however, the accuracy depends on multiple factors, including the type of form deviation, the number of measurement points and their distribution along the circumference, as well as limited access to the full profile [8]. The positioning of measurement points becomes particularly important in the analysis of small angular segments, where the location of the profile apex changes [9]. In such cases, an improper distribution of measurement points can lead to significant distortions in the reconstructed geometry. While most studies have focused on determining the optimal number or density of measurement points, fewer have examined the influence of their spatial location on fitting accuracy [10]. When measuring short arc segments, characteristic points of the profile may be omitted, which can make the obtained results less reliable to interpret. Zhao et al. [11] demonstrated that the length of the analysed arc and the method of data segmentation have a decisive influence on the accuracy of profile reconstruction. Similarly, Zhu et al. [12] addressed the problem of circle fitting to short arc fragments in coordinate measurements, proposing solutions aimed at improving the stability and accuracy of the results. These studies provide a foundation for further investigation into the influence of arc length and its position relative to the profile apex on the accuracy of geometric parameter determination [13].

In roundness evaluation it's possible to choose one of four fitting elements, however each affecting the measurement outcome differently [14]. The Least Squares Circle (LSCI) method is the most commonly applied due to its computational simplicity, stable fitting, and reduced sensitivity to outliers [12]. However, comparative studies have shown that no universal algorithm guarantees the highest accuracy under all conditions, as the method's effectiveness depends on the data characteristics and surface geometry [15]. Mathematical models based on the LSCI approach make it possible to determine the centre and radius of the fitted circle even for interrupted profiles, where the eccentricity is high relative to the part diameter [16]. Zakharov et al. [17] proposed the use of harmonic analysis to compensate for errors resulting from imperfect centring, indicating that the inclusion of harmonic components improves result stability.

According to ISO 1101 [18], roundness deviation is defined as the radial distance between two concentric circles encompassing all measured points of the profile. However, this definition does not always reflect the complex, multi-harmonic nature of real surface deviations. For rotational samples, the deviation distribution along the circumference often exhibits a multi-lobed character, which further complicates evaluation and interpretation [19]. Measurements of roundness, as an example of a coordinate-based measurement technique, can be performed on various devices depending on the required accuracy parameters. These include coordinate measuring machines - both contact [20] and optical [21] - as well as optical scanners [22], measurement scales [23], and even computed tomography systems [24]. However, specialized instruments dedicated solely to the inspection of cylindrical features are often used for this purpose, and these require specific object-adjustment procedures. In measurements performed with specialized form-measuring instruments, factors such as the number of sampled points, the rotational speed of the table, sample diameter, and fixture stability are of crucial importance [25]. Improper alignment of the measured element relative to the rotational axis may introduce systematic errors, and even slight misalignments can distort the results [26]. Therefore, dedicated form-measuring instruments, owing to their superior precision and stability, are preferred in experimental investigations of roundness deviation [11]. Raj et al. [27] emphasized that knowledge of the geometric characteristics of the measured component is essential for ensuring measurement reliability and repeatability, particularly when only a fragment of the profile is analysed.

The present article includes a continuation of the authors' previous simulation-based research, in which the influence of arc length and its position relative to the profile apex on the diameter value of an oval-shaped profile was examined [28]. The current work extends this investigation toward the assessment of form deviation, specifically roundness deviation, under analogous geometric conditions. The measurement of fragmentary profiles represents a significant metrological challenge, particularly in cases where full access to the entire circumference is restricted. For future, this is certainly an area where artificial intelligence will play an increasing role. In geometric metrology, AI will not only support the operator in measurement procedures but also analyze the number of points and their distribution to ensure the most reliable measurement results [29]. The findings presented here contribute to a better understanding of the relationship between the geometry of the measured fragment, its position, and the accuracy of roundness deviation evaluation, forming the basis for developing effective measurement strategies for partial profile assessment.

2 Research Problem and Methodology

Roundness deviation is one of the fundamental form characteristics evaluated in dimensional metrology, providing insight into how faithfully a manufactured rotational part reproduces its nominal circular form. However, its reliable determination depends both on the measurement range and on the method used to fit the measured data to the ideal reference geometry. In industrial practice, roundness measurements often cover only a fragment of the profile, which may lead to a misinterpretation of the actual deviation value.

The research problem addressed in this study concerns the evaluation of the influence of the arc length of the analysed segment and its position relative to the profile on the determination of roundness deviation using the LSCI method. The aim of the investigation was to determine the extent to which the reduction of the measurement range affects the stability of the evaluated deviation and the repeatability of the results, as well as how the geometric complexity of the profile modifies this relationship.

For the experiments, three rings (Fig. 1) with a nominal diameter of 100 mm were used, each characterised by a different type of shape deformation: oval (two-lobed), triangular (three-lobed), and quadrangular (four-lobed).

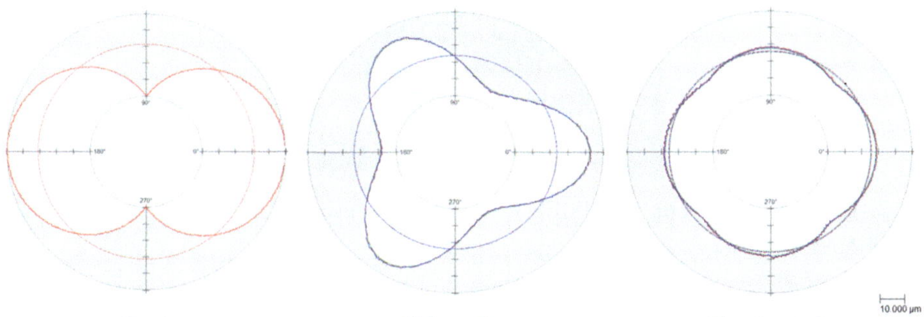

Oval (Roundness deviation: 50.3 µm) | Triangular (Roundness deviation: 37.3 µm) | Quadrangular (Roundness deviation: 9 µm)

Fig. 1. Profiles analysed in the study illustrating different types of form deviation.

The selection of these profiles made it possible to assess how the number of lobes influences the variation of the measurement results obtained for the analysed profiles. Each profile was measured thirty times under repeatability conditions in order to determine the mean roundness deviation and the range of results. The roundness deviation values given in the figure correspond to the mean roundness deviation determined for the complete 360° profile.

The measurements were carried out using a tactile device, the Jenoptik HOMMEL Roundscan 535 (Fig. 2). The analysis included various arc lengths - from the full circumference of 360° to its shortened segments: 270°, 180°, 120°, 105°, 90°, 75°, 60°, 45°, 30°, and 15°. For each case, three positions of the arc segment relative to the profile geometry were considered. These positions correspond to different angular locations of the local extrema of the profile: the peak (0°), where the profile reaches its maximum radial distance; the intermediate position (45°), located between a peak and a valley; and the valley (90°), where the radial distance is the smallest. This distinction reflects how the segment is aligned with respect to the characteristic features of the profile form. Representative outlines of the three analysed profile types are shown in the figures, illustrating the characteristic deformation patterns resulting from the number of peaks and valeys. This approach made it possible to examine the influence of geometric complexity on the accuracy of circle fitting using the LSCI method and to evaluate the stability of results when analysing partial arc segments.

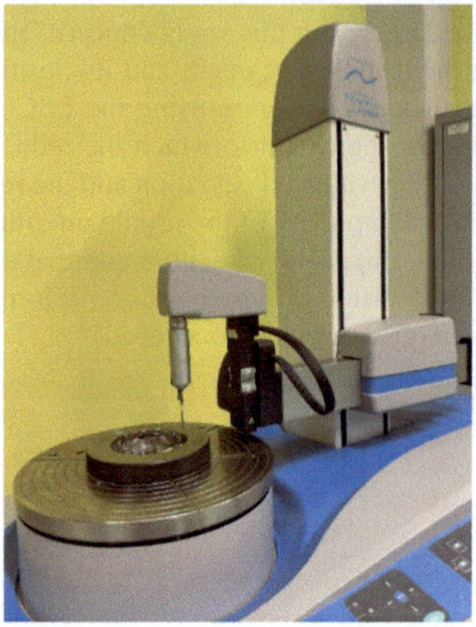

Fig. 2. Form tester Jenoptik HOMMEL Roundscan 535 used for experimental evaluation of roundness deviation.

3 Results

Based on the conducted measurements, datasets were developed that included the values of roundness deviation and the range of results for all analysed arc lengths and segment positions. These results were presented in the form of graphs, allowing the observation of relationships between the length of the analysed segment, the evaluation of roundness deviation, and the repeatability of measurements. The aim of the further analysis is to identify trends, characteristic differences between the individual profile types, and to assess the influence of arc shortening on measurement stability. The graphs and interpretations presented below refer respectively to the oval, triangular, and quadrangular profiles.

3.1 Evaluation for the Oval-Shaped Profile

The first stage of the analysis concerned the oval-shaped profile, which represents the simplest type of form deviation among the tested geometries.

Figure 3 illustrates the relationship between the roundness deviation values of the oval profile and both the length of the analysed arc segment and its position. For the complete contour (360°), the roundness deviation determined using the LSCI method was approximately 50.3 μm on average and was independent of the segment position. However, shortening the analysed arc caused a noticeable decrease in the calculated roundness deviation, the intensity of which depended on the segment's position on the oval. For half of the circumference (180°), the roundness deviation amounted to approximately 2.0 μm for the segment located at the peak of the oval (0°), 31.6 μm for the intermediate position (45°), and 21.1 μm for the valley position (90°). Further reduction of the arc length to 90° and below amplified this effect. At 90°, the mean

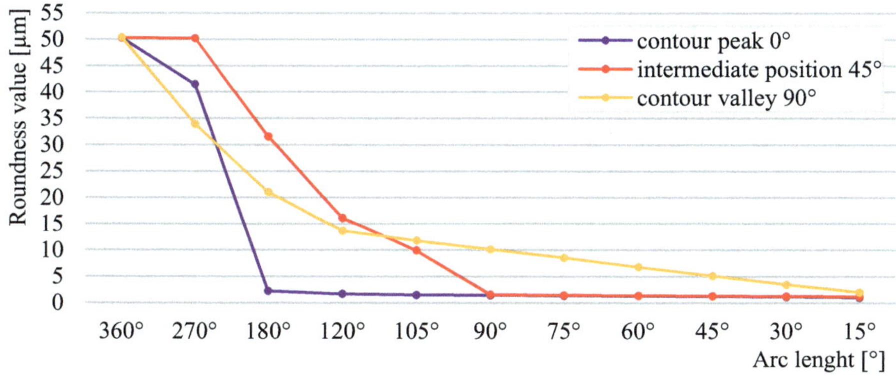

Fig. 3. Mean value of roundness deviation for the oval-shaped profile, depending on arc length and position.

deviation values were approximately 1.4 µm at 0° and 1.5 µm at 45°, while at 90° they reached around 10.0 µm, representing only about 20% of the deviation determined for the full contour. For shorter segments (30° and 15°), the deviations decreased to single micrometers value approximately 1.0 µm at 0° and 45°, and about 2.0 µm at 90°. The trend observed in the graph is unambiguous: as the arc length decreases, the roundness deviation value decreases in a non-linear manner. Initially, the drop is slight in the range from 360° to 270°, then becomes rapid for arc lengths below 180°, reaching near zero values for arcs shorter than 90°. This relationship is particularly evident for the segments located at the peak of the ovality, where the profile is symmetrical and the LSCI-fitted circle aligns almost perfectly with the contour. Segments located diagonally or in the valley region include more asymmetrical portions of the contour; hence, their deviations remain higher, although still significantly underestimated compared to the full profile.

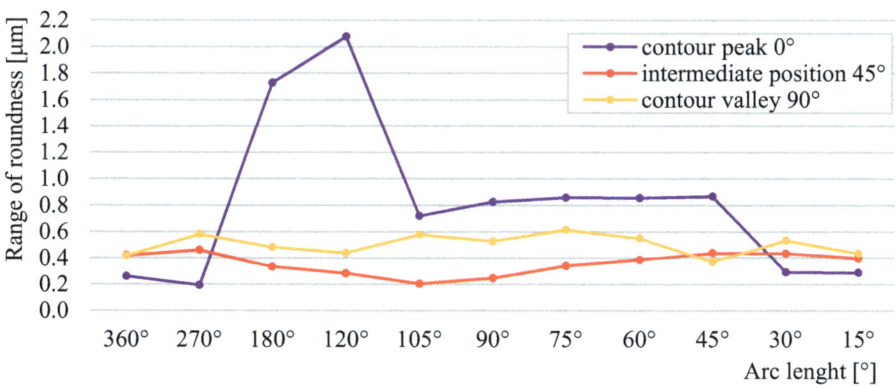

Fig. 4. Range value of roundness deviation for the oval-shaped profile, depending on arc length and position.

Based on thirty repeated measurements of the roundness deviation of the oval-shaped profile, a characteristic distribution of the result range was obtained, reflecting how the measured values change with the length of the analysed arc and the segment's position along the oval profile (Fig. 4). For the complete contour of 360°, the result ranges were relatively small, varying between approximately 0.25 and 0.45 μm, which confirms not only the high repeatability of the full-profile measurement but also the stability of the device when the entire circumference is analysed using the LSCI method. As the arc length decreased, the range did not systematically diminish but instead reached a distinct maximum for the 120° segment at the peak position (0°) of the profile, where it reached approximately 2.1 μm. This indicates the highest variability among all analysed cases. This value clearly exceeded the ranges observed for other positions - both in the valley (90°) and in the intermediate position (45°) - where the maxima did not exceed 0.6 μm. For arc lengths of 180° and 270°, the ranges were also slightly elevated compared with the full circumference, particularly in the valley position (90°), where they reached approximately 0.6 μm. When the arc was further shortened below 90°, the range decreased and stabilised at around 0.4–0.6 μm, with only minor differences between the various segment positions. This indicates that shorter fragments are less sensitive to random variations because they encompass a geometrically less diverse portion of the surface.

3.2 Evaluation for the Triangular-Shaped Profile

The triangular-shaped profile represents a more complex geometric deformation compared to the oval shape, characterised by three evenly distributed peaks and valleys along the circumference. The presence of three distinct curvature extrema leads to greater variation in the local form features, making the evaluated roundness deviation more sensitive to the selected arc length and the position of the measured segment.

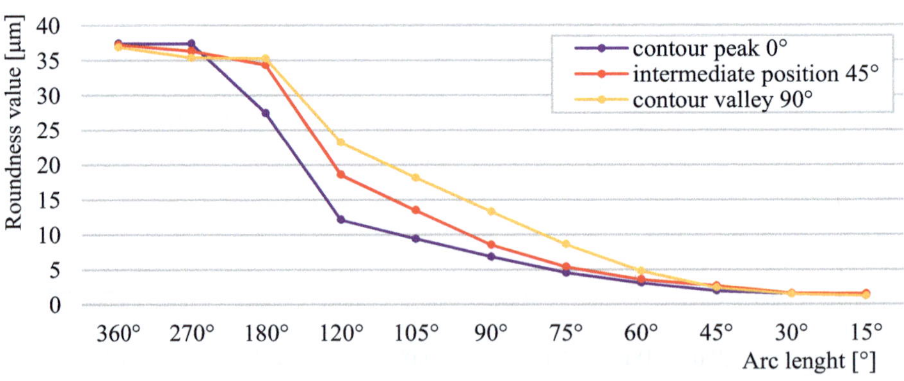

Fig. 5. Mean value of roundness deviation for the triangular-shaped profile, depending on arc length and position.

The analysis of measurement results for the triangular-shaped profile (Fig. 5) indicates significant differences in the trend of roundness deviation values and result ranges compared with the oval-shaped profile. For the complete circumference of 360°, the mean roundness deviation was approximately 37.3 μm, which is lower than in the oval case, while still maintaining a distinct three-lobed shape. As the analysed arc length was reduced, a gradual decrease in the roundness deviation value was observed; however, the reduction was less pronounced than for the oval profile. For an arc length of 270°, the values remained very close to those obtained for the full circumference, ranging from approximately 35.3 to 37.3 μm depending on the segment position. For the 180° segment, a noticeable decrease was recorded, particularly at the peak position (0°), where the mean value dropped to about 27.4 μm. In the intermediate position (45°), the deviation was around 34.3 μm, and in the valley (90°), approximately 35.2 μm. Unlike the oval profile, the triangular profile exhibited a more uniform and gradual change. As the arc length was further shortened, the roundness deviation continued to decrease systematically, but less dynamically than in the oval case. For an arc length of 120°, the mean values ranged from about 12.2 μm at the peak to 23.2 μm in the valley, indicating a position-dependent variation. For 90°, the corresponding values were approximately 6.8 μm at the top, 13.3 μm in the valley, and about 9 μm in the intermediate position. As the arc was further reduced to 60°, 45°, 30°, and 15°, the deviations decreased to 1–3 μm, which follows the same general trend as for the oval profile, although the reduction was less abrupt and the values remained slightly higher for comparable arc lengths.

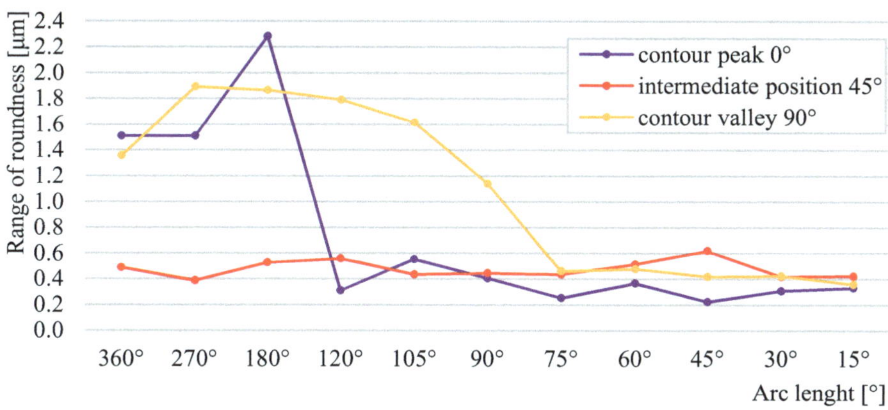

Fig. 6. Range value of roundness deviation for the triangular-shaped profile, depending on arc length and position.

The analysis of the range of roundness deviation values for the triangular-shaped profile (Fig. 6) indicates that, compared with the oval-shaped profile, the scatter of results is generally higher and more strongly dependent on the position of the analysed profile segment. For the full contour of 360°, the range varied from 0.49 μm in the intermediate position (45°) to 1.51 μm at the peak (0°), with 1.35 μm observed in the valley (90°), showing that differences in repeatability between positions are already visible for the complete profile. As the arc length was reduced to 270°, the range at the peak position remained nearly constant, while in the valley it increased markedly, reaching a maximum of 1.89 μm, indicating a deterioration of repeatability in this region. In contrast, at the intermediate position (45°), the range decreased to 0.38 µm, confirming that this setting provides the highest metrological stability. The largest range was recorded for an arc length of 180° at the peak position, where it reached 2.28 µm. This is characteristic of a segment encompassing a single vertex of the triangular profile, where the LSCI method becomes most sensitive to local shape variations. For the same arc length in the valley, the range was lower (1.86 µm), and at the 45° position significantly smaller (0.53 µm), further confirming that intermediate segments exhibit the best repeatability. As the arc length was further shortened to 120° and 105°, the range at the peak position decreased to 0.31–0.55 µm, whereas in the valley it remained relatively high (1.6–1.8 µm), showing that segments located between lobes are more susceptible to local distortions. For arcs shorter than 90°, the range stabilised within a narrow interval of 0.2–0.6 µm across all positions, indicating good repeatability for short arc measurements. Compared with the oval profile, in which the greatest variation occurred for a 120° segment at the top, the triangular profile demonstrates higher stability for shorter arcs but greater differentiation between positions. The intermediate position (45°) remains particularly stable, with the range consistently low (0.38–0.62 µm) regardless of arc length, confirming its advantage for assessing the triangular-shaped profile.

3.3 Evaluation for the Quadrangular-Shaped Profile

The last analysed case concerns the quadrangular-shaped profile, which exhibits the highest geometric complexity among all the examined shapes. The presence of four evenly distributed peaks and valleys results in a more symmetrical distribution of deviations and a lower amplitude of local deformations. The analysis of this profile allows for the assessment of how increased geometric complexity influences the stability and repeatability of roundness deviation measurements.

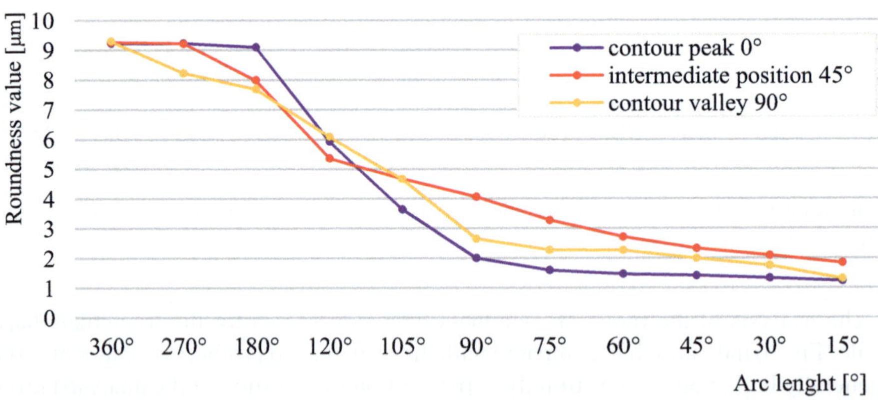

Fig. 7. Mean value of roundness deviation for the quadrangular-shaped profile, depending on arc length and position.

The analysis of the roundness deviation values for the quadrangular-shaped profile revealed a stable and predictable pattern of variation depending on the length of the analysed arc segment and its position relative to the profile (Fig. 7). For the full circumference, the roundness deviation remained at approximately 9 μm, showing no significant differences between the examined segment positions. Reducing the arc length caused a gradual but smooth decrease in deviation values, without the abrupt changes observed in profiles with fewer lobes. For arc lengths of 120° and 90°, the deviation values were within a few micrometres, and for shorter arcs of 30–15°, they decreased to about 1–2 μm, while maintaining only minor differences between positions. It is characteristic that for shorter segments, the highest deviation values appeared at the intermediate position (45°), whereas the lowest occurred at the peak position (0°), indicating the influence of local asymmetry of the segment on the circle fitting result. Overall, the quadrangular profile demonstrates the lowest sensitivity to both arc length and segment position compared to the other analysed profiles. The changes in deviation values occur smoothly and proportionally, confirming the high metrological stability of this contour and its resistance to the effects of limited measurement range.

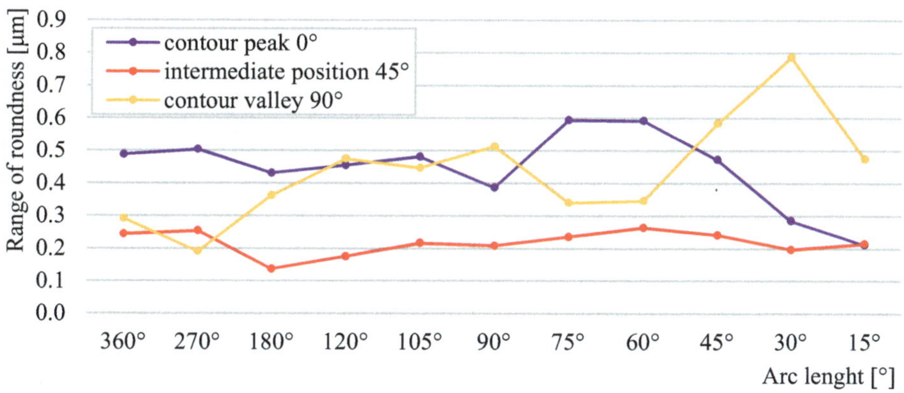

Fig. 8. Range value of roundness deviation for the quadrangular-shaped profile, depending on arc length and position.

The range of roundness deviation values for the quadrangular-shaped profile remains stable across the entire range of arc lengths (Fig. 8), with consistently lower levels than those observed for the oval- and triangular-shaped profiles. This behaviour results from the more uniform distribution of curvature in the four-lobed profile, which reduces the influence of individual peaks or valleys on the LSCI fitting and ensures greater robustness of the measurement. For the full circumference, the range is approximately 0.5 μm at the peak, 0.2 μm at the intermediate position, and 0.3 μm at the valley; when shortened to 270°, these values remain comparable. At 180°, the range stays similarly low, with about 0.4 μm at the peak, 0.1 μm at the intermediate position, and 0.4 μm at the valley, confirming good repeatability even when measuring half of the profile. For shorter arcs, a gentle differentiation between positions is observed: the intermediate position (45°) maintains the lowest and most uniform range, within 0.1–0.3 μm across all lengths; the peak position typically falls within 0.3–0.6 μm; and the valley position shows occasional local increases, reaching up to 0.8 μm for 30° and around 0.5 μm near 90°. Based on these observations, it can be concluded that the quadrangular-shaped profile is the least sensitive to the conditions and strategy of partial measurement, while selecting the intermediate position (45°) further minimizes result variability, whereas the valley position tends to be slightly more susceptible to minor fluctuations for shorter arcs.

4 Discussion

The analysis of the obtained results for the three profile types: oval, triangular, and quadrangular, clearly demonstrates that the number of lobes shaping the profile has a fundamental influence on the behaviour of roundness deviation values and their range, depending on the length of the analysed arc segment and its position on the profile. As the number of lobes increases, both the rate of decrease in deviation values and the nature of measurement repeatability change. The oval-shaped profile shows the highest sensitivity to shortening of the analysed arc. For a half-circumference (180°), the evaluated roundness deviation is significantly reduced compared to the full-profile result, particularly for segments located at the peak position. As the measured arc becomes shorter, this effect intensifies, leading to a loss of reliability in roundness evaluation for partial measurements. In the case of the triangular-shaped profile, the decrease in deviation values with shortening arc length is more gradual but remains dependent on the segment's position relative to the profile. Segments located in the valleys tend to produce higher deviation values than those at the peaks, which results from local curvature variations. The quadrangular-shaped profile, characterised by a greater number of lobes, exhibits the most stable and predictable behaviour. The deviation values decrease steadily with shortening arc length, without abrupt fluctuations, indicating greater metrological stability. With the increasing number of lobes, the form errors become more evenly distributed, and the influence of local deformations on the measurement result diminishes. The analysis of the range values obtained from 30 repeated measurements confirmed the relationship between contour complexity and measurement repeatability. For the oval profile, the range was the highest, particularly at the peak position. The triangular profile showed a more position-dependent scatter - higher in the valleys and lowest at the intermediate position, which provided the greatest stability. The quadrangular profile, in turn, exhibited very low and uniform range values across all analysed arc lengths, confirming its high repeatability regardless of segment geometry. The results clearly indicate that the geometric complexity of the profile plays a crucial role in the metrological stability of the LSCI method, with its robustness increasing alongside the number of lobes in the contour. However, a direct comparison of absolute values between profiles is not justified due to differences in their nominal roundness deviation amplitudes. Therefore, the interpretation of results should focus on observing trends, directions of change, and characteristic behaviours of individual profiles depending on the arc length and position of the analysed segment.

5 Conclusions

The conducted research confirmed that the number of lobes in the profile contour significantly affects the nature of roundness deviation variations and the stability of results obtained using the Least Squares Circle method. As the number of lobes increases, the trend of changes becomes more uniform, and the results exhibit greater repeatability regardless of the length or position of the analysed arc segment. In all analysed cases, shortening the arc length led to a decrease in the measured roundness deviation; however, the rate and intensity of this effect diminished with an increasing number of lobes, indicating higher metrological robustness for profiles with greater geometric complexity. The quadrangular-shaped profile demonstrated the greatest stability and the smallest dispersion of results, whereas the oval-shaped profile showed the highest variability and the greatest sensitivity to the analysed arc length. From a metrological perspective, this means that a higher number of lobes in the profile promotes a more uniform distribution of form errors and improves the circle fitting accuracy of the LSCI method, allowing for reliable results even when only partial profile segments are analysed. The findings also indicate that an appropriately selected arc segment can serve as a representative measurement range, enabling the practical use of limited measurements for rapid shape accuracy assessment of components with complex geometries.

Acknowledgments. The research results presented in this article were obtained thanks to the infrastructure established as part of the project no. POIR.04.02.00-00-D012/20-00 NSMET – National Network of Coordinate Metrology, registered on Polish Road Map of Research Infrastructure, co-funded by the European Regional Development Fund within Smart Growth Operational Programme 2014-2020; 4.2 Development of modern research infrastructure of the science sector.

This research was funded by the Ministry of Science and Higher Education grant no. 0614/SBAD/1603.

The work was supported by the 'PhDBoost' Program for doctoral students of the Doctoral School of Poznan University of Technology (in 2024) from the University's subsidy financed from the funds of Ministry of Science and Higher Education No. 0610/SPHD/0093.

References

1. Grigoriev, S.N., et al.: Analysis of standard un-certainty using the Monte Carlo method for arc measurement on a coordinate measuring machine. Measurement. **253**, 117589 (2025). https://doi.org/10.1016/j.measurement.2025.117589
2. Umare, M., Ambadkar, S.: Estimation of shape factor for irregular particles using three-axial measurement approach. IOP Conf. Ser.: Earth Environ. Sci. (2022)
3. Rhinithaa, P.T., Selvakumar, P., Lawrence, K.D.: Comparative study of roundness evaluation algorithms for coordinate measurement and form data. Precis. Eng. **51**, 458–467 (2018). https://doi.org/10.1016/j.precisioneng.2017.10.001
4. Shen, Y., et al.: Surface form inspection with contact coordinate measurement: a review. Int. J. Extreme Manuf. **5**, 22006 (2023). https://doi.org/10.1088/2631-7990/acc76e
5. Singh, D., Arunachalam, N.A.: Novel iterative field search approach to minimum zone circle for roundness error estimation. J. Comput. Inf. Sci. Eng. **23** (2023). https://doi.org/10.1115/1.4056483

6. Moroni, G., Petrò, S.: Geometric tolerance evaluation: a discussion on minimum zone fitting algorithms. Precis. Eng. **32**(3), 232–237 (2008). https://doi.org/10.1016/j.precisioneng.2007.08.007
7. Gapinski, B., Ocenasova, L., Rucki, M.: Strategy of roundness measurement with CMM. In: Annals of DAAAM and Proceedings of the Inter-National DAAAM Symposium, pp. 527–528 (2008)
8. Li, H., et al.: Uncertainty modeling and evaluation of CMM task oriented measurement based on SVCMM. Measurem Sci Rev. **17**, 226–231 (2017). https://doi.org/10.1515/msr-2017-0027
9. Gapinski, B., Kołodziej, A.: Measurement of diameter and roundness deviation for circle with incomplete contour. In: 11th IMEKO TC14 International Symposium on Measurement and Quality Control, ISMQC, pp. 142–145 (2013)
10. Summerhays, K.D., et al.: Optimizing discrete point sample patterns and measurement data analysis on internal cylindrical surfaces with systematic form deviations. Precis. Eng. **26**, 105–121 (2002). https://doi.org/10.1016/S0141-6359(01)00106-4
11. Zhao, J., et al.: Arc extraction analysis for circularity measurement of small cylindrical parts by the segmenting-stitching method. Instrumentation. **10**, 1–11 (2023). https://doi.org/10.15878/j.cnki.instrumentation.2023.03.001
12. Zhu, J., et al.: Measurement of short arc based on Centre constraint least-square circle fitting. Opt. Precis. Eng. **17**, 2486–2492 (2009)
13. Gapinski, B., Wieczorowski, M.: Measurement of diameter and roundness on incomplete outline of element with three-lobbing deviation. In: Katalinic, B. (ed.) 24th Daaam International Symposium on Intelligent Manufacturing and Automation, 2013. Elsevier Science Bv, Sara (2014)
14. Maresca, P. et al A proposal for the metrological characterization of circular features with digital optical machines. AIP Conference Proceedings 158–168, (2012)
15. Roy, U., Zhang, X.: Development and application of voronoi diagrams in the assessment of roundness error in an industrial environment. Comput. Ind. Eng. **26**(1), 11–26 (1994). https://doi.org/10.1016/0360-8352(94)90024-8
16. Janecki, D., Adamczak, S., Stepien, K.: Calculating associated circle parameters for interrupted roundness profiles. In: Proceedings of the 9th International Symposium on Measurement and Quality Control, ISMQC, pp. 145–149 (2007)
17. Zakharov, O.V., Brzhozovskii, B.M.: Accuracy of centering during measurement by roundness gauges. Meas. Tech. **49**, 1094–1097 (2006). https://doi.org/10.1007/s11018-006-0242-1
18. International Organization for Standardization ISO 1101:2017 Geometrical product specifications (GPS) - Geometrical tolerancing - Tolerances of form, orientation, location and run-out (2017)
19. Gabbia, A., Imkamp, D.: Roundness measurement and its different evaluation methods. In: European Society for Precision Engineering and Nanotechnology, Conference Proceedings - 18th International Conference and Exhibition, EUSPEN, pp. 461–462 (2018)
20. Gapinski, B., Wieczorowski, M., Marciniak-Podsadna, L., et al.: Comparison of different method of measurement geometry using CMM, optical scanner and computed tomography 3D. Procedia Eng. **69**, 255–262 (2014). https://doi.org/10.1016/j.proeng.2014.02.230
21. Grochalski, K., Wieczorowski, M., H'roura. J., Le Goïc, G.: The optical aspect of errors in measurements of surface asperities using the optical profilometry method Front Mech. Eng. 6:12, (2020) 10.3389/fmech.2020.00012
22. Wieczorowski, M., Gapinski, B., Swojak, N.: The use of optical scanner for analysis of surface defects. In: Annals of DAAAM & Proceedings, pp. 76–85 (2019)
23. Swojak, N., Wieczorowski, M., Jakubowicz, M.: Assessment of selected metrological properties of laser triangulation sensors. Measurement. **176**, 109190 (2021). https://doi.org/10.1016/j.measurement.2021.109190

24. Gapiński, B., Wieczorowski, M., Mietliński, P., Mathia, T.G.: Verification of computed tomograph for dimensional measurements. In: International Scientific-Technical Conference Manufacturing, pp. 142–155. Springer International Publishing, Cham (2022)
25. Görög, A., Kuruc, M.: The influence of the setup on the result of measuring the roundness of an Anuloid surface. Machines. **12** (2024). https://doi.org/10.3390/machines12040258
26. Sahay, C., Ghosh, S., Venkatesh, N.A.: An anova based study of variations in circularity form error due to mathematical methods and measuring instrument eccentricity. In: ASME International Mechanical Engineering Congress and Exposition, Proceedings (IMECE) (2023)
27. Raj, A.P., Kumar, R.A., Dhanish, P.B.: Factors influencing the performance of a roundness measuring instrument – development of an empirical model. Measurement (Lond). **131**, 193–203 (2019). https://doi.org/10.1016/j.measurement.2018.08.045
28. Smyczynska, L., Wieczorowski, M., Jakubowicz, M., Gapinski, B.: Simulation of influence of diameter and other circle parameters on results of incomplete round profile testing. Adv. Sci. Technol. Res. J. **19**(7), 151–163 (2025). https://doi.org/10.12913/22998624/203706
29. Wieczorowski, M., et al.: A novel approach to using artificial intelligence in coordinate metrology including nano scale. Measurement. **217**, 113051 (2023). https://doi.org/10.1016/j.measurement.2023.113051

Manual and CMM-Based Inspection Methods for Controlled Radius Evaluation

Adam Szczepański[1](✉) and Marek Magdziak[2]

[1] Pratt & Whitney Rzeszów, Rzeszów, Poland
adam.szczepanski@prattwhitney.com
[2] The Faculty of Mechanical Engineering and Aeronautics, Rzeszów University of Technology, Rzeszów, Poland
marekm@prz.edu.pl

Abstract. Correctly defined design requires strict adherence to appropriate Geometrical Dimensioning and Tolerancing (GD&T) standards. This paper presents an analysis of radius specification based primarily on the ASME Y14.5-2018 standard. According to this standard, a radius can be defined in several ways. Particular attention is given to the distinction between a standard radius (R) and a controlled radius (CR), as well as to the benefits of applying CR to components subjected to high stress. Although the controlled radius offers advantages in many applications, ambiguity persists regarding its quality inspection. In this study, an in-depth analysis of the controlled radius definition was conducted, typical defects were characterized, and key sources of inspection-related uncertainty were identified. Because defects in radius profiles are often too small to be detected with the unaided eye, inspection methods involving microscopes or profile projectors are commonly used in manufacturing. However, these methods are neither efficient nor sufficiently repeatable. Consequently, the concept of using a coordinate measuring machine (CMM) for CR evaluation emerged as a means to automate the process and improve inspection robustness. To assess the suitability of CMM-based inspection, a dedicated artifact was designed to simulate representative defects. The artifact was first evaluated by professional quality inspectors to establish an acceptance-criteria matrix. The radii were then scanned using a CMM, and the measurement results were compared with the matrix, demonstrating the applicability of the machine for CR assessment.

Keywords: Controlled Radius · Quality Inspection · CMM

1 Introduction

The continuously increasing requirements imposed on the mechanical engineering sector have led to more advanced designs, in which every detail of machinery is subject to close scrutiny. The demand for reduced energy consumption and lower fuel usage entails the necessity to optimize resource deployment and enhance efficiency, while ensuring high component strength and resistance to fatigue damage. To meet these demands, numerous design and manufacturing aspects must be taken into account. One of these aspects is the

M. Hryb et al. (Eds.): MANUFACTURING 2026, LNME, pp. 106–120, 2026.
https://doi.org/10.1007/978-3-032-23038-6_8

definition of the radius at transitions between surfaces, which often constitutes a critical area for the reliable functioning of mechanisms. This is particularly relevant for rotating components operating at high speeds, such as electric motors in the automotive industry, aircraft engines, power turbines, and similar applications [1, 2]. The scope of this paper includes:

- a review of the options for defining a radius as standardized in the ISO and ASME specifications, highlighting their respective advantages and limitations,
- the design and analysis of a dedicated artefact containing a set of radii with varying values and simulated defects,
- evaluation of the artifact by professional quality inspectors to develop an acceptance matrix,
- CMM-based evaluation of the artifact to assess the suitability of coordinate metrology for controlled radius inspection.

2 Methods for Specifying Radii

2.1 Radius in the ISO Standard

The Geometrical Product Specification (GPS) system is based on ISO 8015 standard [3], which contains the fundamental principles, ISO 1101 [4] which defines geometrical tolerances, and a set of subject-specific standards that further elaborate these definitions and principles. The ISO standards do not explicitly define the radius, therefore, applying a plus/minus (±) tolerance directly to the radius is generally not recommended (ISO 14405-2, paragraph 7.5 [5]). The preferred approach is to define the radius as a Theoretically Exact Dimension (TED) and to apply a line profile tolerance. However, this method does not necessarily fulfil the designer's intent, as the profile tolerance defines a tolerance zone distributed around the nominal radius, but lacks additional refinements that can fully control the profile shape. Within the acceptable tolerance zone, the profile of the radius may assume any form, provided that it remains continuous. Figure 1 illustrates the contrast between the ambiguous ± radius tolerance and the unambiguous line profile tolerance used to control the radius in the ISO standard.

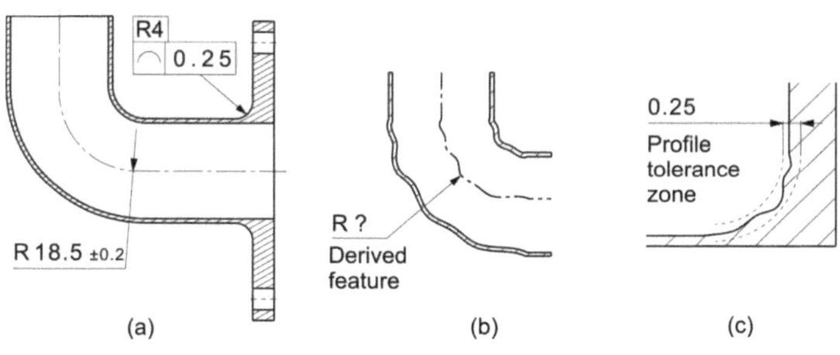

Fig. 1. Radii definition based on the ISO standard: (a) example drawing requirements, (b) R18.5 interpretation, (c) R4 interpretation.

2.2 Radius in the ASME Standard

In contrast to the ISO standards, the ASME standard [6] provides a clear and unambiguous definition of the radius. It distinguishes between the standard radius, denoted as R, and the Controlled Radius, denoted as CR. It is also possible to define a radius as basic dimension (analogous to the TED in the ISO standards) and to control it with a profile of a line requirement. The comparison of the definitions is explained in Fig. 2.

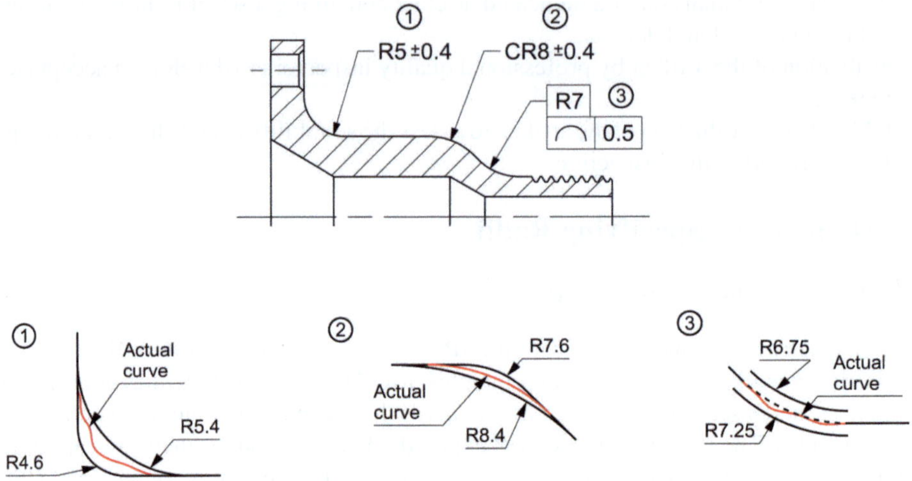

Fig. 2. Radii definition based on the ASME-Y14.5-2018 standard.

According to the standard, when a radius is specified with a bilateral size tolerance and the center is not located, the resulting tolerance zone has a crescent shape. No additional requirements are imposed. Verification of whether the actual curve fits within the tolerance zone is straightforward. However, a common challenge arises when reporting the actual radius value, as the curvature can vary along the feature. Many local radii may be observed, and the locally evaluated radius values may fall outside the tolerance limits, while the curve remains within the crescent-shaped tolerance band.

The Controlled Radius provides improved control of the radius profile, as, in addition to the requirements of the standard radius, further refinements of the feature are defined. The CR must fit within the same tolerance zone as the standard radius, however, the curve must be free of any defects such as flats, reversals, mismatches, cusps, and similar irregularities. Furthermore, no locally observed radius may be smaller or larger than the specified tolerance limits. This type of definition is primarily intended for elements critical to product quality and mechanical reliability. Typical examples include rotating shafts, camshafts, power take-off shaft, discs, and blade hubs. In this paper, this requirement is of particular interest.

The interpretation of a radius controlled by a profile of a line is analogous to the interpretation of a line profile in the ISO standards. The actual radius profile must fit within a tolerance zone obtained by offsetting the nominal curve.

The comparison of radius definitions in both the ISO and the ASME standards is summarized in Table 1. It is fair to say that the ASME standard is more suitable for designs that require precise control to prevent potential stress concentrations localized at corners, fillet radii, and similar features.

Table 1. Comparison of radius definition.

Criteria	ISO	ASME
± R unambiguous definition	No – direct R ± tolerance is not recommended in ISO	Yes – radius has a clear, explicit definition
Profile used for controlling radius	Yes – no additional requirements	Yes – no additional requirements
± Radius size control	Unclear – ISO does not define a size tolerance for R	Yes – radius size can be directly controlled
± R shape control	No – ISO does not provide a dedicated R-form control	Yes – via Controlled Radius

3 CR Investigation

3.1 CR Insight

Although the ASME standard defines the controlled radius in considerable detail, it does not specify any criteria for feature assessment. It is evident that a curve that appears free of defects when inspected with the unaided eye may be deemed unacceptable when examined under magnification. The specification neither defines the size of an unacceptable defect nor prescribes the method of inspection. These aspects should be defined in supplementary company documents, as stated in [6], notes to paragraph 5.16.2. Examples of unacceptable defects are presented in Fig. 3.

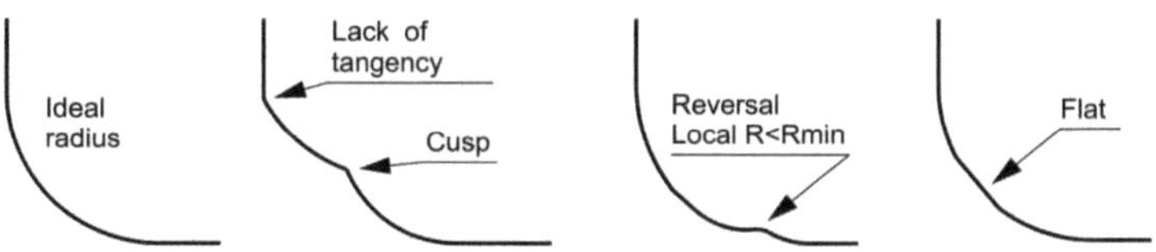

Fig. 3. Example of ideal radius and typical defects.

Examples of defects found on actual manufactured components produced by a turning process are shown in the set of images in Fig. 4. Because the evaluated radii were small (< 2 mm), a profile projector equipped with a 20× magnification lens was used for observation. Due to the fact that the radii were located on a large component, positioning the part directly on the projector's stage was not feasible. Consequently, a dedicated mold was fabricated to obtain a replica, which was subsequently sectioned into slices for further examination. This approach enables a detailed assessment of the enlarged radii, however, it requires substantial manual effort, which may affect the quality assessment process. The main factors influencing inspection accuracy include mold deformations, non-perpendicularity of the slices relative to the radius direction, and discrepancies in visual evaluation.

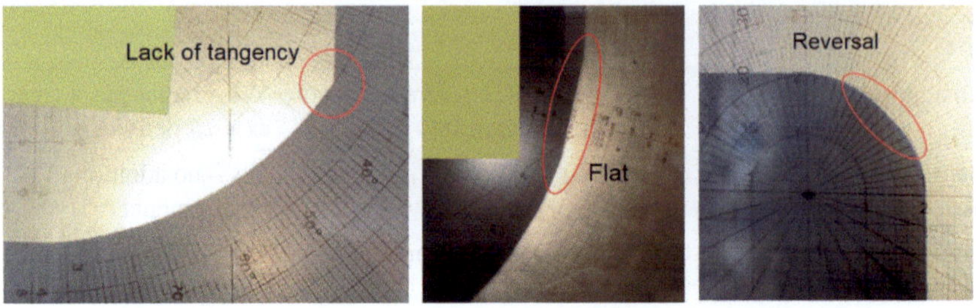

Fig. 4. Examples of defective radii observed with a profile projector.

3.2 An Artifact

In order to evaluate the feasibility of using a CMM for assessing the controlled radius and its capability to detect radius defects, a dedicated artifact containing radii of various sizes with intentionally introduced defects was designed and manufactured. Figure 5 presents the overall view of the artifact and includes an example of the applied defect for the 12th radius. As shown in the images in Fig. 4, in some cases the defect may be difficult to identify visually. Therefore, an additional purpose of the reference object is to standardize the evaluation procedure and calibrate the inspectors' sensitivity.

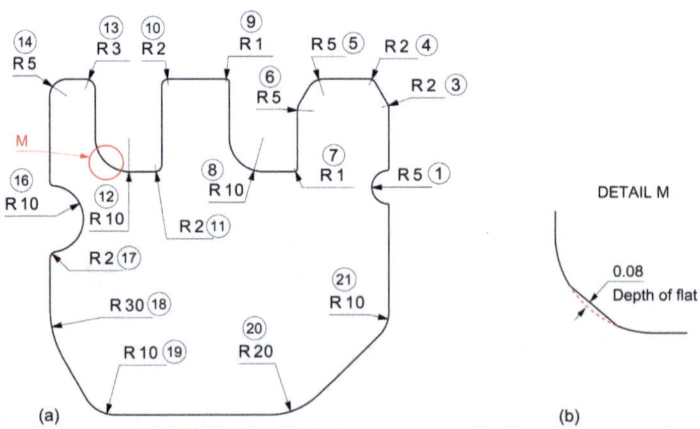

Fig. 5. The artefact: (a) overall view, (b) magnified detail M.

The artefact consists of a flat, comb-shaped sheet metal plate containing a set of radii: 2 mm, 5 mm, 10 mm, and 20 mm, each including a controlled defect in the form of a flattened area. The defect depth increases progressively from a very small value of 0.005 mm up to 0.08 mm. The nominal radii and corresponding defect parameters are listed in Table 2. Based on the authors' experience, the sequence of defects was designed to cover the most probable detection limit observed in industrial practice. The artefact was manufactured using the wire electrical discharge machining (wire EDM) method.

Table 2. List of radii and corresponding defects.

No.	R [mm]	DOF [mm]	LEN [mm]	No.	R [mm]	DOF [mm]	LEN [mm]
3	R2	0.005	0.28	14	R5	0.03	1.1
4	R2	0.01	0.4	8	R10	0.005	0.63
10	R2	0.015	0.49	8	R10	0.01	0.89
10	R2	0.02	0.56	8	R10	0.015	1.1
11	R2	0.05	0.9	8	R10	0.02	1.26
5	R5	0.005	0.45	8	R10	0.03	1.55
5	R5	0.01	0.63	12	R10	0.05	2
5	R5	0.015	0.77	12	R10	0.08	2.5
14	R5	0.02	0.89	20	R20	0.02	1.79
14	R5	0.025	1	20	R20	0.05	2.83

Notes: No. is the designator of a radius; R – the radius value; DOF – the depth of the defect; LEN – the length of the flat area.

The relationship between DOF and LEN as a function of radius size is presented in Fig. 6. Identifying which of these parameters drives the acceptance criterion is relevant for the subsequent stages of research. The further studies presented in this paper are based on DOF analysis.

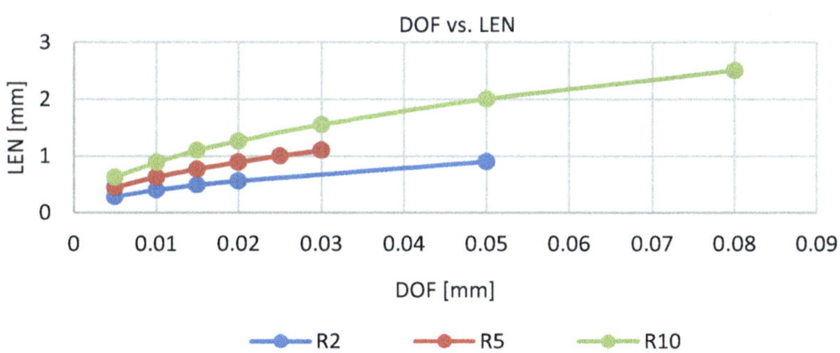

Fig. 6. Relationship between DOF and LEN depending on radius.

3.3 Inspection of the Artifact

In order to evaluate the correctness of the template manufacturing process and to assess the feasibility of using the selected measurement system for controlled radius evaluation, a CMM inspection was performed. Factors influencing CMM-based inspection relevant to controlled radius evaluation such as measurement point distribution, scanning speed, calculation algorithms, measurement data filtration and the influence of probe radius compensation have been comprehensively analyzed in [7] and the references cited therein. In the case of filtration, the effect of data-point filtering as a function of the low-pass filter cutoff frequency is summarized in Table 2 of [7]. Building on these findings, the present work assumes compliant measurement conditions and focuses on defect detectability and decision consistency rather than on the performance limits of the measurement system.

The profile of reference object was scanned using the Mitutoyo Crysta-Apex series machine equipped with the PH10MQ articulating head, the SP25 scanning probe with the SM1 module, and the stylus with a 0.5 mm ruby ball. The accuracy of the coordinate measuring system, according to the manufacturer's specifications, are as follows:

- $E_{0,MPE} = \pm 1.7 + 4\ L/1000\ \mu m$
- $R_{0,MPL}$ – limit not specified by the manufacturer (actual value: 0.6 μm)
- $P_{[Form.Sph.Scan:PP:Tact,MPE]} = 2.3\ \mu m$ (actual value: 1.3)

The scan resolution was set to 0.02 mm, which resulted in approximately 47,000 data points. Such point density was required to ensure sufficient sensitivity for detecting even the smallest defects. The 0.5 mm probe size was selected to ensure that the intentional defects are not mechanically filtered out, which can occur when using larger ball diameters. To minimize scanning noise and eliminate outliers, a robust spline filter defined in ISO 16610-32 [8] was applied, with the wavelength setting of 0.2 mm. The results of comparing the actual artifact profile with the nominal model are presented in Fig. 7, which, as an example, shows the upper portion of the template along with two enlarged representative radii.

The applied profile tolerance is 0.02 mm (±0.01 mm), represented in the figure by the tolerance band between the green curves. The dark curve shows the actual profile, which fits within the tolerance band, and the labels indicate the maximum and minimum local deviations. The actual scan was best-fitted to the nominal profile using the Gaussian method. The chart shows that the assumed manufacturing tolerance was achieved in the EDM process, and all simulated defects are clearly identifiable. Therefore, the inspection results will be used to verify whether the coordinate measuring machine is capable of detecting defects at the same level as a profile projector with a 20× magnification lens.

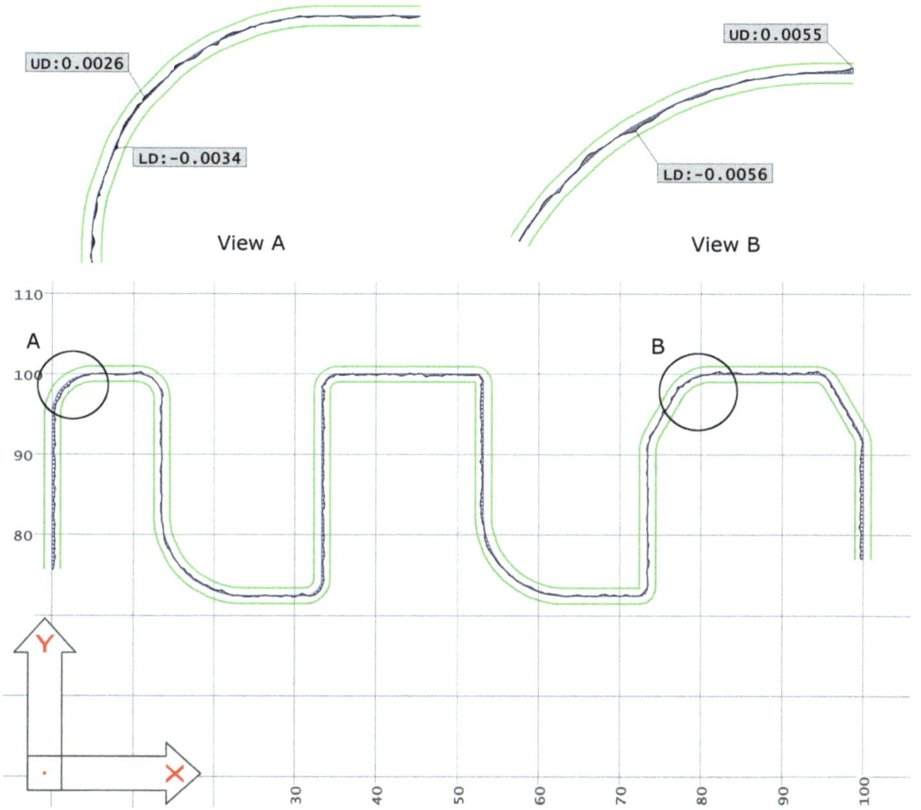

Fig. 7. Actual artifact profile verification with respect to the nominal model (CMM).

As an independent verification of the radius profile, the MarSurf CD25 profilometer equipped with the contour probe (tip radius 0.025 mm, tip angle 30°) was used. The manufacturer-declared parameters include a guideway deviation of 0.4 µm over 140 mm and a contacting error of 0.75 µm. The scan resolution was set to 0.001 mm, and the scanning speed was 0.5 mm/s. As it was not feasible to scan the entire contour of the artifact in a single setup, each radius was measured independently. The software supplied with the device does not provide a direct comparison between the measured and nominal profiles. Therefore, the acquired data were exported and analyzed using the CMM software. The results of the profile tolerance evaluation are presented in Fig. 8 for the same two radii indicated in Fig. 7. As can be observed, the deviation differences between the CMM and the profilometer do not exceed 2.3 µm, which is sufficient for the purposes of this study.

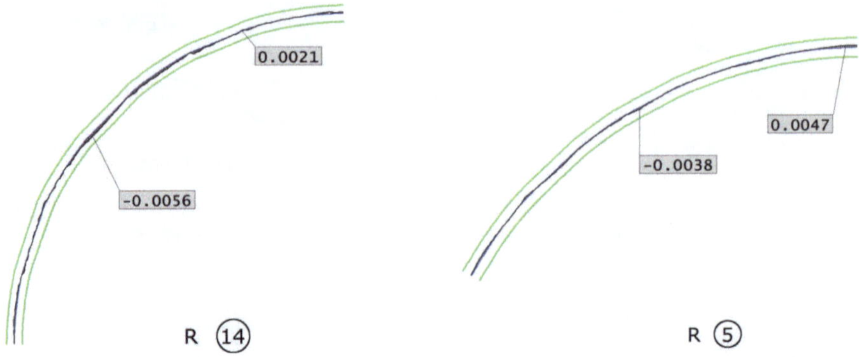

Fig. 8. Actual artifact profile verification with respect to the nominal model (profilometer).

3.4 Reproducibility Analysis

The standard manual method for evaluating small radii on manufactured components involves creating the replica and performing the visual assessment using a profile projector. Human factors introduce variability in defect detection and conformity decisions. Demonstrating that a CMM can reliably identify defects with accuracy comparable to a 20× profile projector would therefore enhance the repeatability and objectivity of quality control. In this analysis, experienced inspectors from a manufacturing company, relying on their expertise, provided a pass/fail judgment for each radius on the artifact. Since the artifact's size allowed it to be placed directly on the stage of the profile projector, the observation and inspection process was simplified. Because the radii were small, a 20× magnification lens was used to enlarge them on the projector screen. Inspectors were instructed to evaluate each radius and, in cases of uncertainty, to classify the feature as being at the boundary of the acceptance level. The evaluation criteria were defined as follows:

- defect non-detectable – score 0,
- defect detectable – score 1,
- defect at the detection boundary– score 0.5.

Based on the assessments, the matrices presented in Tables 3, 4, and 5 were created and subsequently used for further analysis. Overall, the inspectors' evaluations were consistent. However, a few cases showed minor differences in judgment. These differences did not contradict each other but rather reflected borderline situations in which a defect could be considered either acceptable or unacceptable. There were no instances in which inspectors had directly opposing opinions.

Table 3. Evaluation chart for R2.

	0.005	0.01	0.015	0.02	0.05
Inspector 1	0.0	0.5	1	1	1
Inspector 2	0.0	1	1	1	1
Inspector 3	0.0	1	1	1	1

DOF [mm]

Table 4. Evaluation chart for R5.

	0.005	0.01	0.015	0.02	0.025	0.03
Inspector 1	0.0	0.5	1	1	1	1
Inspector 2	0.0	1	1	1	1	1
Inspector 3	0.0	0.5	1	1	1	1

DOF [mm]

Table 5. Evaluation chart for R10.

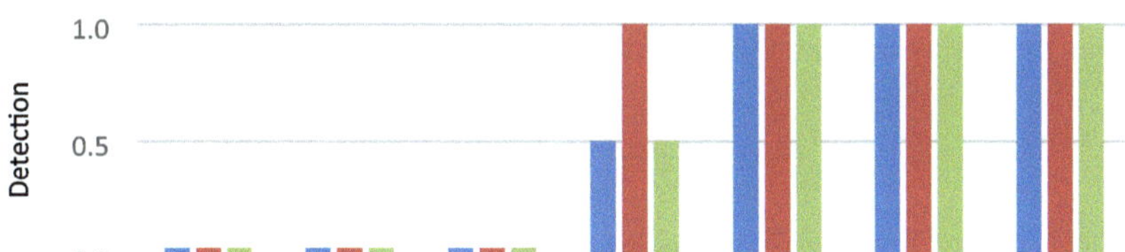

	0.005	0.01	0.015	0.02	0.03	0.05	0.08
Inspector 1	0.0	0.0	0.0	0.5	1	1	1
Inspector 2	0.0	0.0	0.0	1	1	1	1
Inspector 3	0.0	0.0	0.0	0.5	1	1	1

DOF [mm]

Inter-rater agreement was assessed using Fleiss' kappa for three nominal categories (0, 0.5, 1). The obtained overall value $\kappa = 0.785$ indicates good agreement among the three experts, confirming acceptable reproducibility of the attribute-based inspection process. Category-specific analysis revealed excellent agreement for categories 0 and 1, while the intermediate category (0.5) showed substantially lower agreement, reflecting expert uncertainty rather than measurement inconsistency. The kappa statistic quantifies inter-rater agreement beyond chance, while its standard error, Z-statistic, and associated p-value assess the statistical reliability and significance of the observed agreement. The parameters were calculated using Minitab statistical software and they are presented in Table 6.

Table 6. Fleiss' Kappa Statistics.

Response	Kappa	SE Kappa	Z	P(vs > 0)
0	1.000	0.136	7.348	0.000
0,5	0.339	0.136	2.489	0.006
1	0.762	0.136	5.598	0.000
Overall	0.785	0.109	7.228	0.000

Based on the evaluation, the detection limits were determined:

- for radius R = 2 mm, the detection threshold is at DOF = 0.01 mm, as two of the three evaluations were positive,
- for radius R = 5 mm, the detection threshold is at DOF = 0.015 mm, since at 0.01 mm only one evaluation was positive,
- for radius R = 10 mm, the detection threshold is at DOF = 0.03 mm, since at 0.02 mm only one evaluation was positive,
- for radius R = 20 mm, only two defects were simulated, as this study focuses primarily on radii up to 10 mm. For the first defect at DOF = 0.02 mm, all evaluations were negative (score 0), while for the second defect at DOF = 0.05 mm, all inspectors provided positive evaluations (score 1).

Following the experiment, the acceptance criteria can be estimated for the entire radius range from R = 2 mm to R = 10 mm. As shown in Fig. 9, the relationship between the radius size and the detection level can be approximated with a linear model.

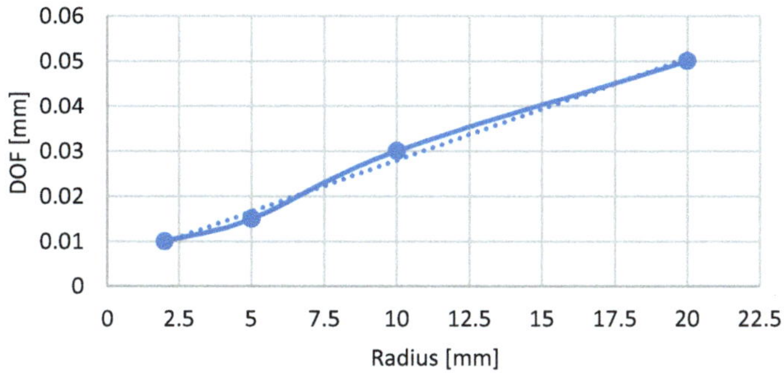

Fig. 9. Relationship between radius size and detectable defect depth.

3.5 Visual Versus CMM Assessment

Having evaluated the reference object and established the detection matrices, the next step was to verify whether the same defects could be identified in the CMM-scanned profile. Considering that the length of flattening at a given defect depth varies significantly with radius size (see Table 2), while the detection level based on DOF can be approximated by a linear model, the analysis focused primarily on the defect depth. The study was conducted in a continuous and coherent manner, encompassing the artifact design, its evaluation by quality inspectors, and the measurement data acquired using a coordinate measuring machine. As an illustrative example, one of the radii (R = 2 mm, radius No. 10) is presented in Fig. 10. This comparison demonstrates the steps leading to the validation of the coordinate measuring machine as a suitable tool for accurately representing actual component defects.

Radius No. 10 was designed with two defects having depths of 0.015 mm and 0.02 mm. The inspectors' evaluations confirmed that both defects were detectable and that their decisions were consistent. The inspected area is shown in Fig. 10(c), captured using a profile projector with 20× magnification. The arrows indicate the defects corresponding to the designed features shown in Fig. 10(b). Subsequent figures illustrate the curve acquired through scanning. Figure 10(d) presents the complete profile of the radius No. 10, with the flattened regions circled. Figure 10(e) and (f) show magnified views of the 0.02 mm and 0.015 mm defects, respectively. Based on the scanned points, a reference circle, depicted as *r10(1)*, was constructed, excluding the points located at the flat regions. The defect depth was then determined as the distance between the reference circle and the defect profile. The actual measured values are shown in the labels *Distance*. As observed, the measured depths closely match the designed values: the 0.015 mm defect was measured as 0.0152 mm, and the 0.02 mm defect as 0.0196 mm. The analysis was performed using MCOSMOS software.

These results confirm the suitability of CMM measurement as a method capable of accurately reflecting the actual condition of the component, thus providing a foundation for further research on the application of this method in industry.

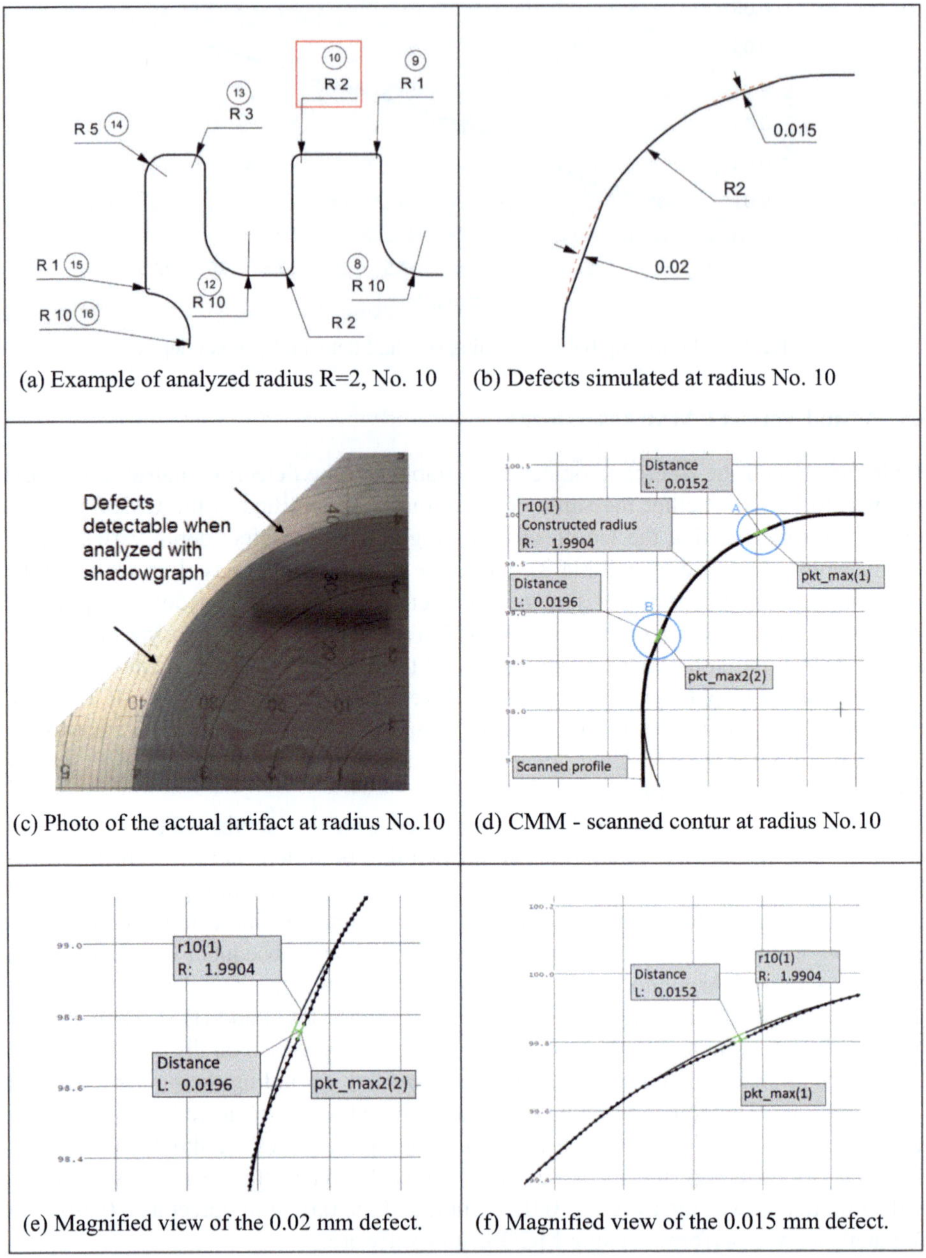

(a) Example of analyzed radius R=2, No. 10

(b) Defects simulated at radius No. 10

(c) Photo of the actual artifact at radius No.10

(d) CMM - scanned contur at radius No.10

(e) Magnified view of the 0.02 mm defect.

(f) Magnified view of the 0.015 mm defect.

Fig. 10. Comparison of nominal geometry and actual radius profile, including defect detection using a profile projector and CMM data.

4 Summary

The article presents the development and validation of a dedicated reference artifact designed for evaluating the capability of coordinate measuring machines to detect small geometrical defects on radii of manufactured components. A high-resolution CMM scan was performed to verify the reference object manufacturing accuracy, which was also validated with a profilometer. The results confirmed that all simulated defects were produced as intended and the deviation did not exceed 6 μm. A reproducibility study was then conducted in which experienced inspectors evaluated each radius using a profile projector with 20× magnification. Their assessments were largely consistent, and any differences reflected borderline cases rather than contradictions. Subsequently, the visual evaluations were compared with the data acquired from coordinate measurements. Each radius and the flattened regions localized on it were analyzed, confirming that the defects detected using the profile projector were also identifiable in the data obtained from CMM scanning. As an illustrative example, radius No. 10 demonstrated that both defects (0.015 mm and 0.02 mm) were clearly detectable in the CMM scan. The measured defect depths closely matched the designed values, confirming the measuring system's ability to represent small geometric deviations at a level comparable to visual inspection. The study concludes that coordinate measuring machines are capable of reliably detecting defects of the same order of magnitude as those identified using profile projectors, thereby reducing dependence on subjective human evaluation and improving the robustness and effectiveness of quality control processes for precision-manufactured components. Future work will focus on the automation of controlled radius conformance assessment based on coordinate measurement data.

References

1. Kienzle, N., Hoang, D.H.M., Waesker, M., Buelten, B., di Mare, F., Doetsch, C.: Influence of fillet radii on the flow and strength behavior of a shrouded centrifugal compressor impeller. In: Proceedings of the 14th European Conference on Turbomachinery – Fluid Dynamics & Thermodynamics (ETC14). European Turbomachinery Society (Euroturbo), Gdańsk (2021)
2. Abbas, A.W.N., Hussein, A., Al Shubber, H.A.: A numerical analysis of shaft stress concentration using ANSYS. Eng. Technol. Appl. Sci. Res. **15**(5), 26282–26289 (2025)
3. International Organization for Standardization (ISO): ISO 8015:2011 – Geometrical product specifications (GPS) – Fundamentals – Concepts, principles and rules. International Organization for Standardization, Geneva (2011)
4. International Organization for Standardization (ISO): ISO 1101:2017 – Geometrical product specifications (GPS) – Geometrical tolerancing – Tolerances of form, orientation, location and run-out. International Organization for Standardization, Geneva (2017)
5. International Organization for Standardization (ISO): ISO 14405-2:2018 – Geometrical product specifications (GPS) – Dimensional tolerancing – Part 2: Dimensions other than linear or angular sizes. Geneva (2018)
6. The American Society of Mechanical Engineers (ASME): ASME Y14.5-2018: Dimensioning and Tolerancing: Engineering Product Definition and Related Document Practices. ASME, New York (2018)

7. Magdziak, M.: Application of coordinate measuring machines for analysis of a controlled radius based on linear regression. Adv. Sci. Technol. Res. J. **18**(2), 11–25 (2024)
8. International Organization for Standardization (ISO): ISO 16610-32:2015 – Geometrical product specifications (GPS) – Filtration – Part 32: Spline filters. Geneva (2015)

Analysis and Optimization of Manufacturing Processes and Tool Wear

Evaluation of Surface Topography, Microstructure and Tool Wear After Turning Inconel 718 Alloy with a Tool Overhang of 10 × D

Krzysztof Smak[1,2](✉), Piotr Szablewski[2,3], and Stanisław Legutko[1]

[1] Faculty of Mechanical Engineering, Poznan University of Technology, Poznan, Poland
krzysztof.smak@prattwhitney.com
[2] Pratt & Whitney Kalisz, Kalisz, Poland
[3] The President Stanislaw Wojciechowski Calisia University, Kalisz, Poland

Abstract. The work concerns the evaluation of the edge wear and surface quality, defined by the topography and microstructure of the surface layer of the machined surface, using a 10 × D tool holder overhang. The tests were carried out on samples made of Inconel 718 material, in the conditions of finishing turning with constant cutting parameters. The specimens were subjected to heat treatment, consisting of solution heat treatment and precipitation hardening, as a result of which a hardness of 45 ± 2 HRC was obtained. A turning holder with a vibration damper with a diameter of 70 mm and an overhang of 700 mm was used. This type of holder is dedicated to turning internal surfaces. The tests were carried out using V-shaped cutting inserts from five different tool manufacturers, made of coated fine-grained cemented carbides. The values of the obtained surface topography parameters and the microstructure of the surface layer were assessed. The edge wear was also evaluated. Based on the conducted tests, it was found that the surface roughness, as well as the microstructure of the surface layer, is mostly influenced by the wear of the edge. For the cutting edge that exhibited the highest wear, as assessed by the VB_C indicator, the highest values of surface topography parameters were obtained, amounting to $Sa = 2.38 \pm 0.026$ μm and $Sz = 12.1 \pm 1.12$ μm at the preliminary phase of machining and $Sa = 4.37 \pm 0.037$ μm and $Sz = 15.7 \pm 0.95$ μm in the terminal phase of machining.

Keywords: Inconel 718 · Turning · Surface Topography · Microstructure of the Surface Layer

1 Introduction

Inconel 718 is a material belonging to the group of nickel-based heat-resistant alloys. Owing to its mechanical and physical properties, such as high resistance to corrosion and creep, as well as its ability to maintain the required strength and hardness at elevated temperatures in the range of 800–900 °C, it is widely used in the aerospace industry for the production of hot section components of turbine engines [1, 2]. The characteristic

M. Hryb et al. (Eds.): MANUFACTURING 2026, LNME, pp. 123–136, 2026.
https://doi.org/10.1007/978-3-032-23038-6_9

features of alloy 718 also include its low thermal conductivity [3], thermal diffusivity, and high work-hardening tendency; therefore, it is considered as a difficult-to-machine material [4, 5]. One of the groups of aircraft engine components manufactured from Inconel 718 alloy are main shafts. This kind of component, due to its high slenderness ratio, presents additional challenges during the machining process. These difficulties are associated with the necessity of using long-overhang tools to machine internal surfaces. Considering the finishing machining of this material, most studies focus on tool wear and the integrity of the resulting surface [6, 7]. Hua and Liu [8] conducted dry turning tests of Inconel 718. The study showed that changes in cutting speed do not have a significant effect on surface roughness. Studies on the influence of cutting speed on surface roughness during turning of Inconel 718 were also carried out by D'Addona et al. [9] and Jafarian and Umbrello [10]. The authors concluded that increasing the cutting speed initially improves surface roughness however, after exceeding a certain value, surface quality deteriorates, which they attributed to rapid wear of the cutting edge. Bushlya et al. [11] investigated the influence of tool wear and machining parameters on the intensity of white layer formation during machining of Inconel 718. The authors' experimental study led to the conclusion that tool wear, dry machining, and cutting speed were the main factors contributing to the intensification of the white layer. Yazid et al. [12], based on their conducted studies, concluded that the microstructural slip angle and the depth of its influence are greater under dry machining conditions. It is also worth noting that the majority of studies are conducted using standard turning tool holders, which ensure sufficient stiffness of the tool–workpiece–fixture system .

The aim of this study is to investigate the effect of tool wear in a tool assembly with a $10 \times D$ overhang on the surface topography and the microstructure of the surface layer after the finish turning of Inconel 718 alloy. The research is conducted to improve the machining process of the inner surface of an aircraft engine shaft–type component.

2 Material and Methods

The experiments were carried out on an M40 Multitask center manufactured by WFL. The material adopted in the research was aged Inconel 718 alloy with a hardness of 45 ± 2 HRC, prepared in the form of a sleeve. The samples used in the study had the following dimensions: OD of 132 mm, ID of 77.4 mm, and length of 122 mm. V-shaped cemented carbide cutting inserts were tested (Table 1). The experiment for each cutting edge was carried out three times; therefore, three specimens were machined with each cutting edge. An anti-vibration tool holder with a diameter of 70 mm (S-570-3C-70/80-1053-40CR, Sandvik) was used, coupled with an SL-SVLBL-40-16HP head with a 700 mm overhang (Fig. 1). The cutting parameters were selected based on previous studies (Table 2). The cutting fluid pressure was set at 80 bar. The research was conducted using longitudinal turning of the inner diameter.

Table 1. List of cutting inserts used in the study.

No.	Designation	Manufacturer	Type of coating	Clearance angle α [°]	Radius of the cutting edge rounding, r_n [μm]
1	VBGT 160408–UM S205	Sandvik	CVD	5	35.127
2	VCGT 160408–MM4 WSM01	Walter	PVD	7	12.44
3	VBGT 160408LF KCS10B	Kennametal	PVD	5	29.944
4	VBMT 160408–LS MP9005	Mitsubishi	PVD	5	27.045
5	VCMT 160408–SM IC804	Iscar	PVD	7	52.865

This study focused on tests aimed at developing a technological solution enabling the optimization of internal turning of a shaft-type part, which requires the use of a tool with a 10 × D overhang. Therefore, for each insert, 14 passes were performed, with a length corresponding to the required spiral cutting distance of a single tool pass, amounting to 1620 m. Each cutting edge has machined material for 1110 s. For each tool, four measurements of the resulting surface topography and the microstructural properties of the surface layer were performed after specific passes, representing control points along the machined surface corresponding to the cutting distance of the tool.

The required surface roughness for the machined surface was $Ra = 1.6$ μm. The machining strategy was planned so that, after achieving the cutting distance for a given control point, subsequent turning operations were carried out on a shortened section enabling the assessment of surface quality. The surface topography was evaluated using the *Sa* and *Sz* parameters, measured at the aforementioned control points. For each sample, measurements were taken in three locations along the diameter, spaced at 120°, using a Nanoscan 855 device from Jenoptik (Hommel – Etamic) Company. The surface topography parameters were measured using a Gaussian filter on a sampling area of 1.2 mm × 1 mm, in accordance with ISO 25178. Standard deviations were calculated for the recorded measurement results. Microstructure images were captured using a Nikon EPIPHOT 200 microscope at 500× magnification. The samples were embedded in Duracryl and etched with Kalling's reagent after polishing. Tool wear was evaluated based on EDS analysis and SEM images of the used cutting edges, obtained using a

Fig. 1. The tool used in the research: 1) prismatic adapter, 2) anti-vibration tool holder, 3) boring head.

scanning electron microscope (Mira 3, Tescan, Czech Republic) equipped with EDS (Oxford Instruments Ultim Max 65, UK), in accordance with ISO 3685.

Table 2. Cutting parameters adopted in the study.

Cutting conditions	Cutting speed v_c [m/min]	Feed rate f [mm/rev.]	Depth of cut a_p [mm]
Value of parameters	85	0.14	0.2

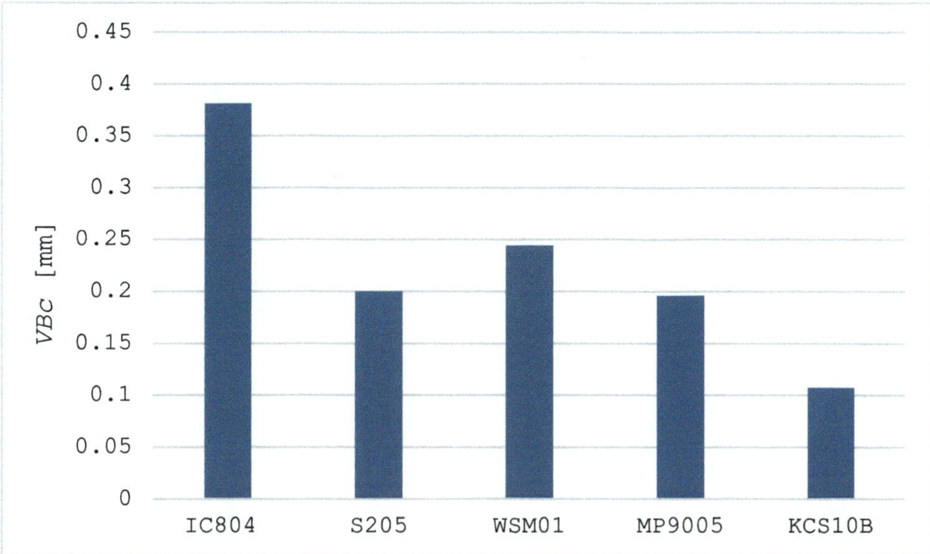

Fig. 2. Tool wear values for cutting edges operating in a 700 mm overhang holder, as defined by the VB_C index.

3 Results and Discussion

Tool wear was evaluated using the VB_C indicator, which defines the width of the flank wear land, as well as based on EDS analysis and SEM images of the cutting edges. It should be noted that each cutting edge machined for 1110 s, which is a relatively long machining time for Inconel 718 alloy. The highest value of the flank wear width index was recorded for the IC804 insert, whereas the lowest value was observed for the KCS10B insert (Fig. 2).

In the case of the IC804 insert, severe cutting edge degradation occurred due to abrasive and strength-related tool wear [13, 14]. The insert exhibited significant cutting edge loss, with chipping and edge breakouts observed [15]. The cutting edge was also characterized by the presence of a significant built-up edge, resulting from adhesive wear [16, 17]. Peeling and oxidation of the coating were also observed on the rake and flank faces (Fig. 3). Additionally, grooves formed on the rake face due to the flowing chip.

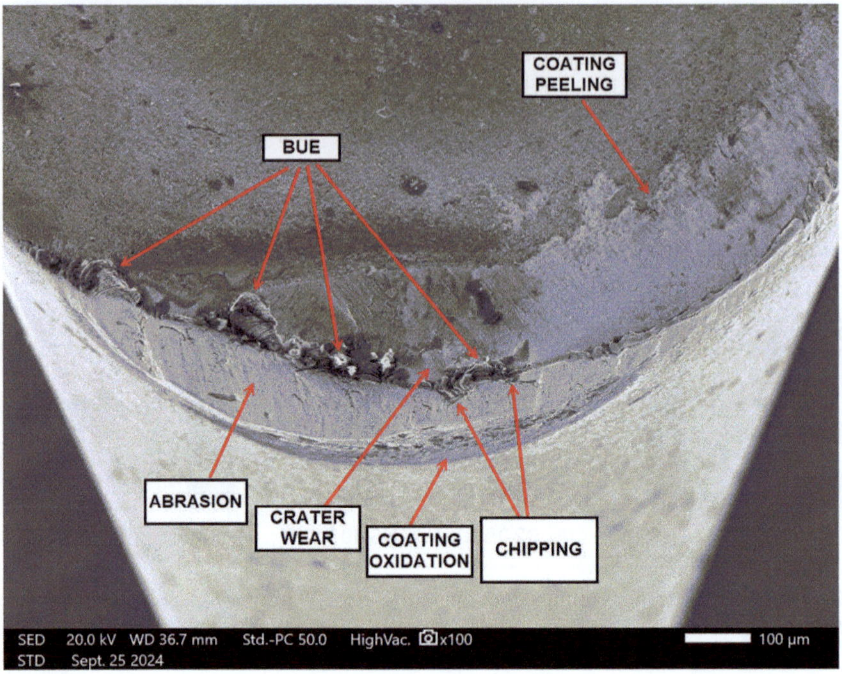

Fig. 3. SEM view of the cutting edge of the worn IC804 insert.

In the case of the KCS10B insert, based on the VB_C parameter, its cutting edge exhibited high strength and wear resistance. Minor abrasive and strength-related tool wear in the form of chipping was also observed, though to a much lesser extent. A groove also appeared on the rake face (Fig. 4). The cutting edge of the KCS10B insert also exhibited a significantly lower tendency to form a built-up edge. Coating oxidation and peeling, combined with delamination, were observed, particularly on the rake face. It should be noted that the KCS10B insert exhibited the lowest wear value, as assessed by the VB_C parameter, which was attributed to the applied PVD AlTiN coating, known for its high wear resistance. This finding is corroborated by the results reported by Zhang et al. [18].

With regard to surface topography, the maximum values of the Sa and Sz parameters were evaluated for the first control point, when the tool was at the initial stage of wear, and for the last control point, when the tool had undergone significant wear. Analyzing the surface topography obtained with the IC804 insert, it can be observed that the tool experienced rapid wear from the very beginning of the operation, as the topography parameter values were relatively high. For the surface under discussion, the recorded values were $Sa = 2.38 \pm 0.026$ μm i $Sz = 12.1 \pm 1.12$ μm (Fig. 5).

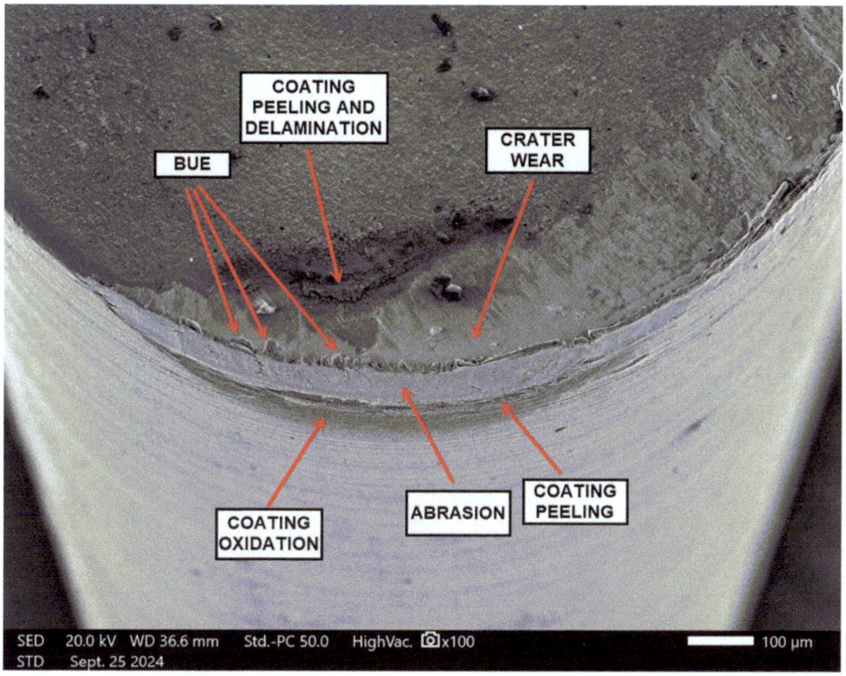

Fig. 4. SEM view of the cutting edge of the worn KCS10B insert.

Focusing on the surface topography at the last control point, it can be observed that significant tool wear, combined with the reduced stiffness of the tool holder, resulted in a considerable increase in surface stereometry parameters.

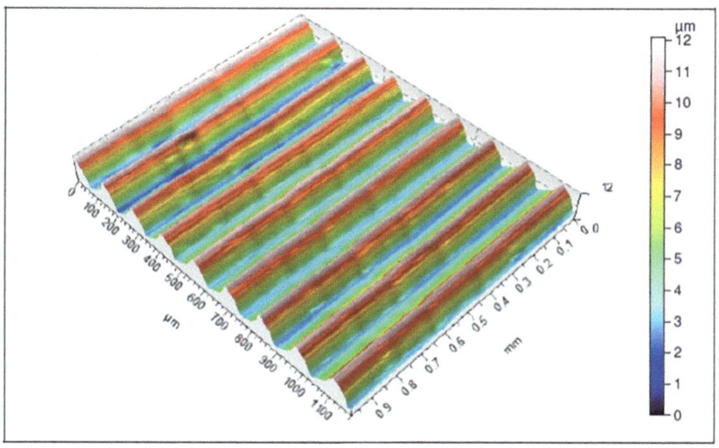

Fig. 5. Surface topography obtained with new IC804 cutting edge.

Their values were $Sa = 4.37 \pm 0.037$ μm and $Sz = 15.7 \pm 0.95$ μm, respectively (Fig. 6). In addition to the relatively high topography parameter values, the surface also exhibited irregularities [19, 20].

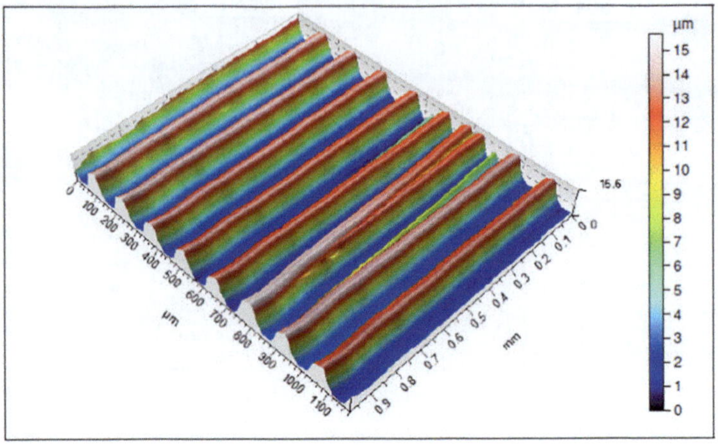

Fig. 6. Surface topography obtained with worn IC804 cutting edge.

Considering the surface topography obtained during the initial stage of machining with the KCS10B insert, it can be concluded that the tool did not experience significant wear at this stage, as the surface stereometry parameter values did not show a substantial increase relative to the programmed machining parameters. For the mentioned surface, the recorded values were $Sa = 1.06 \pm 0.02$ μm i $Sz = 8.71 \pm 1.34$ μm (Fig. 7).

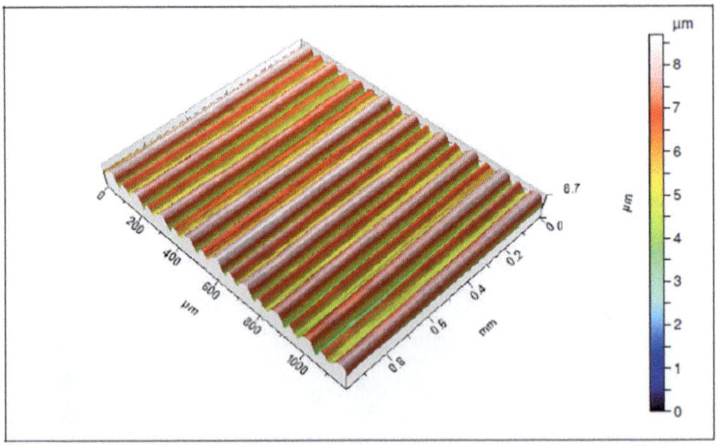

Fig. 7. Surface topography obtained with new KCS10B cutting edge.

Despite the relatively low *Sa* and *Sz* parameter values, certain surface irregularities can also be observed, which may be attributed to the reduced rigidity of the tool holder. In the case of the surface topography obtained with the KCS10B insert at the final stage of machining, the influence of tool wear on the magnitude of the *Sa* and *Sz* parameters is evident. For the surface under discussion, the parameter values were recorded at the level of $Sa = 1.43 \pm 0.18$ µm i $Sz = 9.41 \pm 1.32$ µm (Fig. 8).

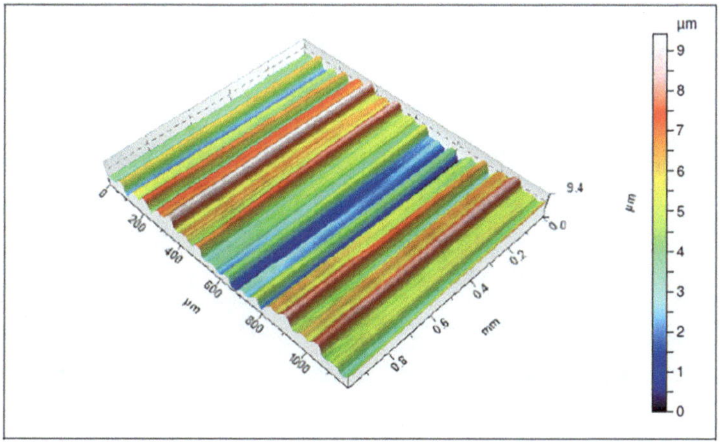

Fig. 8. Surface topography obtained with worn KCS10B cutting edge.

It should be noted that, in addition to the increased topography parameter values, the surface profile exhibited even greater irregularity compared to the surface obtained at the preliminary stage of machining. This can be attributed to the increased tool wear and the resulting cutting forces, which reduced the stability of the boring bar during operation.

In evaluating the microstructure of the surface layer, surfaces obtained at both the preliminary and terminal stages of machining with the aforementioned inserts were also considered [11]. In the case of the surface layer structure of the surface machined with the IC804 insert at the initial stage of machining, severe grain deformations are visible, reaching a depth of up to 3.8 µm (Fig. 9).

The pronounced grain orientation indicates a rapidly progressing tool wear process even at the initial stage of machining. This deformation extends along the entire length of the examined specimen, suggesting that the material was not properly cut even at this stage. With regard to the surface layer microstructure at the final stage of machining, severe grain deformation was also observed, although its depth reached 2.5 µm (Fig. 10). For the surface in question, pronounced grain orientation was also observed, although the depth of this deformation was smaller. The reduction in deformation depth was most likely caused by a weakened cutting process and an increased sliding effect of the tool over the material. This is the result of significant tool wear, as well as the deflection of the anti-vibration boring bar caused by increased cutting forces.

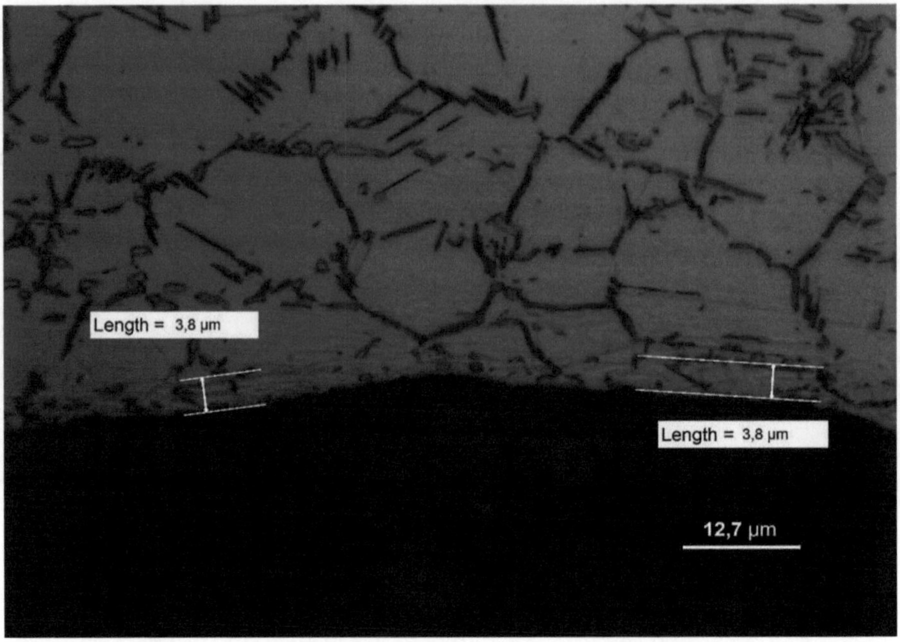

Fig. 9. Surface layer microstructure after turning with new IC804 cutting edge.

By examining the surface layer microstructure of the surface machined with the KCS10B insert, it can be clearly concluded that this surface is free from deformation, indicating that the cutting process proceeded properly and the tool retained its cutting properties (Fig. 11). No grain orientation is observed in the subsurface layer. Analysis of the surface layer obtained at the terminal stage of machining also shows that this layer contains no deformation or significant grain orientation (Fig. 12). This indicates that the cutting process proceeded correctly, which is confirmed by the tool wear, assessed using the VB_C parameter, showing the lowest value among the tested inserts [21, 22].

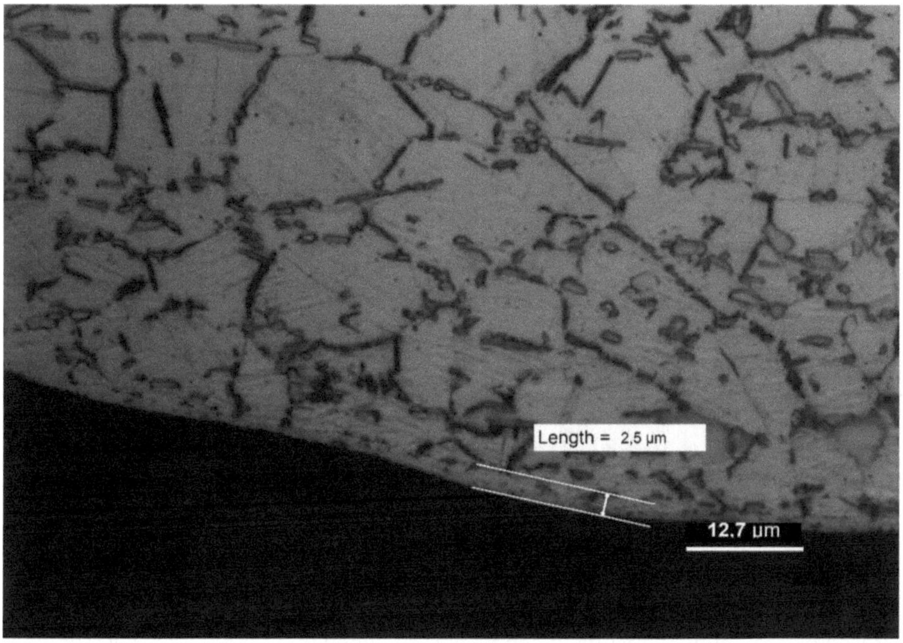

Fig. 10. Surface layer microstructure after turning with worn IC804 cutting edge.

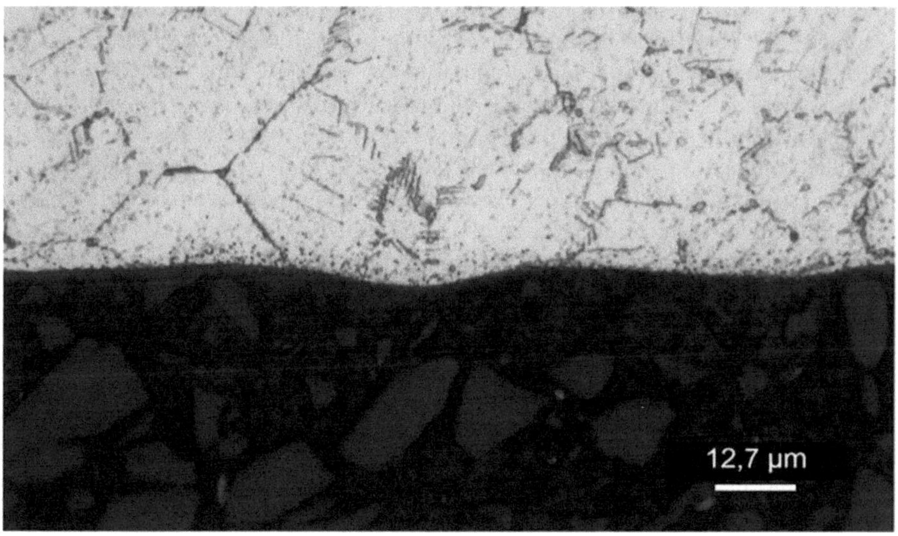

Fig. 11. Surface layer microstructure after turning with new KCS10B cutting edge.

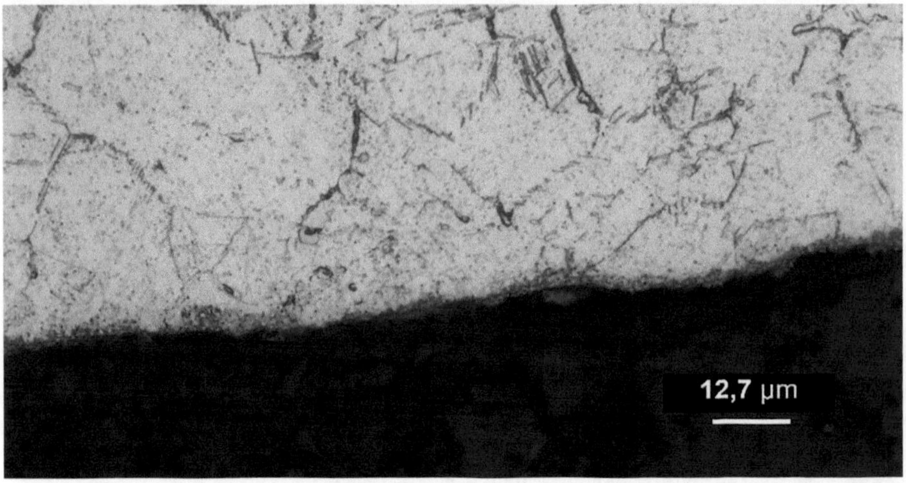

Fig. 12. Surface layer microstructure after turning with worn KCS10B cutting edge.

4 Conclusions

1. Based on the obtained results, it can be clearly stated that tool wear has a significant impact on surface topography and the microstructure of the surface layer, especially when the tool operates in a long-overhang holder.
2. For the IC804 insert, severe abrasive wear of the flank face and significant cutting edge damage were observed. This insert also exhibited a high tendency to form a built-up edge. For this edge the wear index value was $VB_C = 0.381$ mm. The KCS10B edge, on the other hand, showed less wear, and its wear index value was $VB_C = 0.107$ mm.
3. Evaluation of the surface topography shows a strong correlation between its profile and the obtained *Sa* and *Sz* parameter values with tool wear. Combined with the reduced stiffness of the tool holder over the machining time, this led to a deterioration in the quality of the machined surface. This is evidenced by the very high values of the *Sa* and *Sz* parameters recorded on the machined surface in the terminal stage of machining with the IC804 cutting edge, the value of which was $Sa = 4.37 \pm 0.037$ μm and $Sz = 15.7 \pm 0.95$ μm. While for the surface obtained after machining with the KCS10B edge, which exhibited lower wear, the values of topography parameters obtained at the terminal stage of machining were at the level of $Sa = 1.43 \pm 0.18$ μm and $Sz = 9.41 \pm 1.32$ μm.
4. Analysis of the surface layer microstructure images indicates that the relatively rapid wear of the IC804 insert caused grain deformation already at the initial stage of machining, which was not observed in the surface layer condition of the material machined with the KCS10B insert.
5. Analysis of the obtained surface layer microstructure of the machined surface produced with the IC804 insert, at both the preliminary and terminal stages of machining, revealed that despite increasing tool wear and deterioration of surface topography, no increase in the depth of grain deformation was observed. This is a very interesting result that requires further investigation.

References

1. Grzesik, W., Niesłony, P., Habrat, W., Sieniawski, J., Laskowski, P.: Investigation of tool wear in the turning of Inconel 718 superalloy in terms of process performance and productivity enhancement. Tribol. Int. **118**, 337–346 (2018)
2. Smak, K., Szablewski, P., Legutko, S., Petru, J., Kratochvil, J., Wencel, S.: Evaluation of the influence of the tool set overhang on the tool wear and surface quality in the process of finish turning of the Inconel 718 alloy. Materials. **17**(18), 4465-1-4465-31 (2024)
3. Diaz-Alvarez, J., Tapetado, A., Vazquez, C., Miguelez, H.: Temperature measurement and numerical prediction in machining Inconel 718. Sensors. **17**, 1531 (2017). https://doi.org/10.3390/s17071531
4. Szablewski, P.: Evaluation of the topography and load capacity of cylindrical surfaces shaped in the process of finish turning of the Inconel 718 alloy. Measurement. **223**, 113749 (2023)
5. Xavior, M.A., Manohar, M., Patil, M.M., Jeyapandiarajan, P.: Investigation of surface integrity during turning Inconel 718. Trans. Can. Soc. Mech. Eng. **41**(3) (2017)
6. Szablewski, P.: Tribological behavior of Inconel 718 alloy when cutting with a carbide insert prepared using the SPS technique under dry and lubricated sliding conditions. Tribol. Int. **199**, 109951 (2024)
7. Krawczyk, B., Szablewski, P., Legutko, S., Smak, K., Gapiński, B.: Anomalies in the Geometric surface structure of shaped elements composed of Inconel 718 alloy. Materials. **14**, 7524 (2021). https://doi.org/10.3390/ma14247524
8. Hua, Y., Liu, Z.: Effects of cutting parameters and tool nose radius on surface roughness and work hardening during dry turning Inconel 718. Int. J. Adv. Manuf. Technol. **96**, 2421–2430 (2018). https://doi.org/10.1007/s00170-018-1721-7
9. D'Addona, D.M., Raykar, S.J., Narke, M.M.: High speed machining of Inconel 718: tool wear and surface roughness analysis. Proc. CIRP. **62**, 269–274 (2017)
10. Jafarian, F., Umbrello, D., Golpayegani, S., Darake, Z.: Experimental investigation to optimize tool life and surface roughness in Inconel 718 machining. Mater. Manuf. Process. **31**(13), 1683–1691 (2016). https://doi.org/10.1080/10426914.2015.1090592
11. Bushlya, V., Zhou, J.M., Lenrick, F., Avdovic, P., Stahl, J.-E.: Characterization of white layer generated when turning aged Inconel 718. Procedia Eng. **19**, 60–66 (2011)
12. Yazid, M.Z.A., Cheharon, C.H., Ghani, J.A., Ibrahim, G.A., Said, A.Y.M.: Surface integrity of Inconel 718 when finish turning with PVD coated carbide tool under MQL. Procedia Eng. **19**, 396–401 (2011)
13. Lotfi, M., Jahanbakhsh, M., Akhawan, F.A.: Wear estimation of ceramic and coated carbide tools in turning of Inconel 625: 3D FE analysis. Tribol. Int. **99**, 107–116 (2016)
14. Hao, Z.P., Fan, Y., Lin, J., Yu, Z.: Wear characteristics and wear control method of PVD-coated carbide tool in turning Inconel 718. Int. J. Adv. Manuf. Technol. **78**, 1329–1336 (2015)
15. Khochtali, H., Ayed, Y., Zemzemi, F., Bensalem, W.: Tool wear characteristics in rough turning of Inconel 718 with coated carbide tool under conventional and high-pressure coolant supplies. Int. J. Adv. Manuf. Technol. **114**, 2371–2386 (2021)
16. Cakiroglu, R.: Machinability analysis of Inconel 718 Superalloy with AlTiN-coated carbide tool under different cutting environments. Arab. J. Sci. Eng. (2021). https://doi.org/10.1007/s13369-021-05626-3
17. Zhou, J., Bushlya, V., Avdovic, P., Stahl, J.E.: Study of surface quality in high speed turning of Inconel 718 with uncoated and coated CBN tools. Int. J. Adv. Manuf. Technol. **58**, 141–151 (2012). https://doi.org/10.1007/s00170-011-3374-7
18. Zhang, B., Njora, M.J., Sato, Y.: High-speed turning of Inconel 718 by using TiAlN and (Al, Ti) N coated carbide tools. Int. J. Adv. Manuf. Technol. **96**, 2141–2147 (2018)

19. Wieczorowski, M., Bigerelle, M., Brown, C., Pawlus, P., Reizer, R., Pereira, A.: Surface inspection and description in metrology and tribology—Vol.1. Materials. **15**, 5636 (2022)
20. Krolczyk, G., Kacalak, W., Wieczorowski, M.: 3D parametric and nonparametric description of surface topography in manufacturing processes. Materials. **14**, 1987 (2021). https://doi.org/10.3390/ma14081987
21. Thakur, D.G., Ramamoorthy, B., Vijayaraghavan, L.: Some investigations on high speed dry machining of aerospace material Inconel 718 using multicoated carbide inserts. Mater. Manuf. Process. **27**, 1066–1072 (2012)
22. Smak, K., Szablewski, P., Legutko, S., Krawczyk, B., Miko, E.: Investigation of the influence of anti-wear coatings on the surface quality and dimensional accuracy during finish turning of the Inconel 718 alloy. Materials. **16**(1), 80-1-80-13 (2023)

Effect of the Tool Lead Angle on the Effectiveness of Surface Defect Removal Using Ceramic Brushes

Jakub Matuszak(✉), Krzysztof Ciecieląg, Agnieszka Skoczylas, and Kazimierz Zaleski

Mechanical Engineering Faculty, Department of Production Engineering, Lublin University of Technology, Lublin, Poland
j.matuszak@pollub.pl

Abstract. The paper reports findings from a study evaluating the efficiency of ceramic brushes in removing surface defects. Surface defects may appear in the form of scratches, grooves, or burn marks, and their characteristics depend on the machining method applied to the workpiece. Damage to machine components typically initiates at the surface or within subsurface layers, and surface defects can act as stress concentrators, leading to crack propagation. In the experimental procedure, surface scratches of two specified depths were created using custom-designed scratch generators. To remove these defects, dedicated tools for surface defect and burr removal were used, equipped with two different types of ceramic fiber bundles. Four values of the tool lead angle were applied: 0°, 15°, 30°, and 45°. The tool lead angle The tool lead angle is usually set by programming the machining strategy on multi-axis machine tools. The study demonstrated that, There is an optimal lead angle value that ensures effective removal of surface defects. Surface roughness after processing was also analyzed.

Keywords: Surface Defect · Ceramic Brush · Lead Angle

1 Introduction

Lightweight aluminum alloys find widespread use in the aerospace and automotive sectors owing to their high strength relative to density. Components made from these alloys are subjected to advanced machining on multi-axis CNC machine. In the case of large-sized parts with complex final geometries, up to 90% of the material is removed as chips, typically through milling. During part unclamping from the machine, transportation, or storage, surface defects such as scratches may occur. These defects can adversely affect the manufactured component, not only in terms of appearance but also in terms of operational efficiency, as surface defects may act as stress concentrators. Therefore, it is important to apply methods capable of restoring the original condition of the components.

Defects generated during the manufacturing process and following operation of a product may vary significantly in form, depth, shape or distribution on the part surface, as

M. Hryb et al. (Eds.): MANUFACTURING 2026, LNME, pp. 137–146, 2026.
https://doi.org/10.1007/978-3-032-23038-6_10

the conditions under which they occur can differ considerably. Among the most prevalent surface defects are scratches, cracks, grooves, burn marks, pits, craters and burrs [1, 2]. In turn, the authors of studies [3, 4] described the elimination of defects occurring in wood-based structural materials. The properties of the surface layer are divided into physical and stereometric. Surface defects are classified as part of the stereometric features of the subsurface region. Burnishing and shot peening can be an efficient method for changing the properties of the surface layer using cold work [5–8]. The authors [9] presents an analysis of the effectiveness of slide burnishing in eliminating surface defects.

One of the simple and cost-effective methods for removing surface defects, with the potential of automation, is brushing process. Rotating brushes with flexible filaments make it possible to achieve a targeted surface finish. The most common applications of brushing include deburring, surface polishing and cleaning, edge rounding, as well as the removal of surface defects. Eliminating surface defects may additionally produce a characteristic brushed texture, which is often used for decorative purposes. Brushing affects the surface free energy and can be applied as a pre-treatment before bonding or to achieve specific surface layer properties [10]. Filaments striking the machined surface result in impact energy that produces surface layer characteristics similar to shot peening, burnishing, and polishing processes [11, 12]. Surface defects can also be identified by analyzing the recorded force signal Moreover, brushes are well suited for automated machining processes, as the flexible filament tips naturally adapt to the machined surface, removing the need for complex positioning or tool measurement systems [16–19]. Brushes are widely used due to their high performance, suitability for both manual and automated surface defect removal, and the uniform distribution of cutting forces across the workpiece surface [20–22]. This, in turn, enables precise material removal, easier workpiece setup, and minimizes potential damage to both machine and workpiece.

This study investigates the influence of the tool lead angle on the efficiency of surface defect removal, focusing on groove-type defects produced with specialized surface defect generators. The experiment used three different lead angle settings and compared with conventional machining, in which the tool axis is perpendicular to the machined surface. Two defect depths and two types of fibers, differing in their intensity of interaction with the machined surface, were used.

2 Methodology

A schematic representation of the experimental setup is presented in Fig. 1. The experiments were conducted on samples of EN AW 2024 aluminum alloy. Cuboidal plates with starting dimensions of 15 × 8 × 80 mm were prepared as samples. The milling and brushing were carried out on a Dematec DEM4000 CNC machine. The prepared samples were subjected to face milling, maintaining constant parameters ($v_c = 500$ m/min; $a_p = 0.5$ mm; $f_z = 0.1$ mm/tooth). After milling the samples were subjected to the process of generating groove-type surface defects. To remove surface defects, brushing was performed using brushes with replaceable ceramic fibers.

Surface topography measurements following brushing were taken using a T8000RC120-400 profilographometer provided by Jenoptik Hommel-Etamic (Germany, Jena, Villingen-Schwenningen,). The surface topography examination was performed on the 4.8 mm × 4.8 mm surfaces. Visual inspection was carried out with the use of a Keyence VHX-500 microscope with software enabling the measurement of surface defect depth.

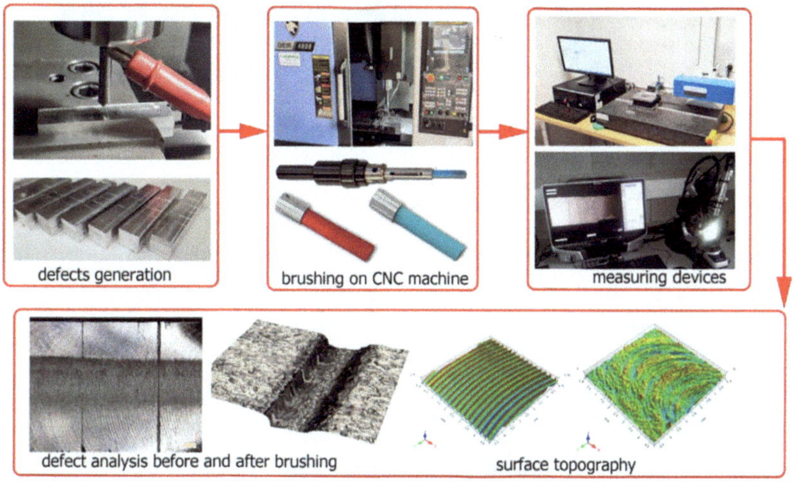

Fig. 1. Experimental set-up and procedure

2.1 Formation of Surface Defects

Surface defects, scratches perpendicular to the feed direction of milling and brushing were produced on the surfaces of the tested samples. To generate the scratches, special styluses, referred to as surface defect generators, were used. These tools had a corner rounding radius of approximately 40 μm and were manufactured from a cemented carbide rod. Grinding was performed on the working surfaces to produce a conical geometry, with an apex angle of 90°. Figure 2 shows the defect generators and an image of the stylus tip, along with the measurement of the rounding radius.

Surface defects were produced on a Dematec DEM4000 machine using pre-prepared defect generators in combination with a digital multimeter. Repeatability of the scratch size was achieved by employing a short-circuit detection function activated upon contact between the scratch generator and the workpiece. A multimeter probe was attached to the scratches generator, while the other was connected to the prismatic sample. The generator was moved toward the sample surface in 10 μm increments to determine the point at which a short circuit occurred. After detecting the short circuit, the stylus was indented into the surface by a predefined depth. Two nominal defect depths were used in the study: 20 μm and 60 μm. The feed of the generator was set to 300 mm/min. The length of the grooves was 15 mm (the entire width of the sample). Figure 3 shows the measurement setup used for producing surface defects.

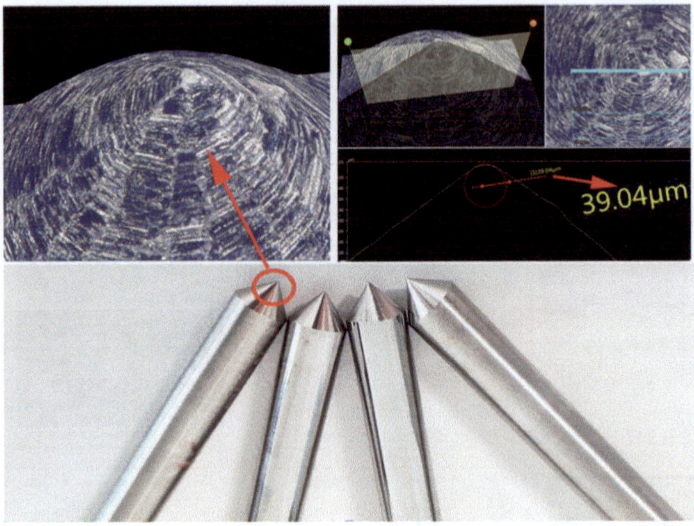

Fig. 2. Defect generator view and rounding radius measurement

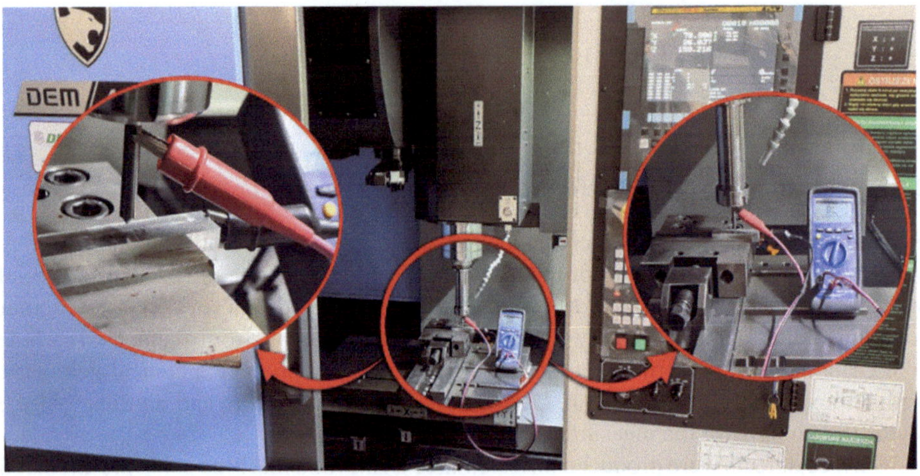

Fig. 3. Setup for generating surface defects

2.2 Brushing Process

Xebec brushes were used for the removal of surface defects (Fig. 4). The sleeve was positioned in a specialized floating holder, with ceramic fibers inside that could extend from 5 to 30 mm. The tests employed two ceramic fiber types: red, which is flexible and designed for soft materials and blue with the high stiffness dedicated to machining difficult-to-cut materials.

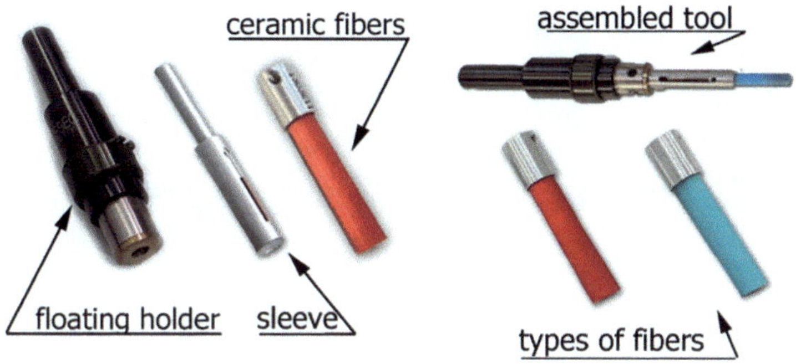

Fig. 4. Xebec tool design

Constant brushing parameters were applied: a feed-in of 0.5 mm, a spindle speed of n = 8000 rpm, and a feed rate of 300 mm/min. The fiber projection from the sleeve was set to 5 mm.

2.3 Lead Angle and Brushing Setup

During multi-axis machining, the tilt angle represents rotation around the feed axis (F) and was fixed at 0° throughout the tests, while the lead angle corresponds to rotation of the tool around the cross feed axis (C) was varied in the experiment in the range of 0 to 45°. These angles define the tool's orientation, which affects the cutting speed, tool wear, and the efficiency of removing surface defects (Fig. 5).

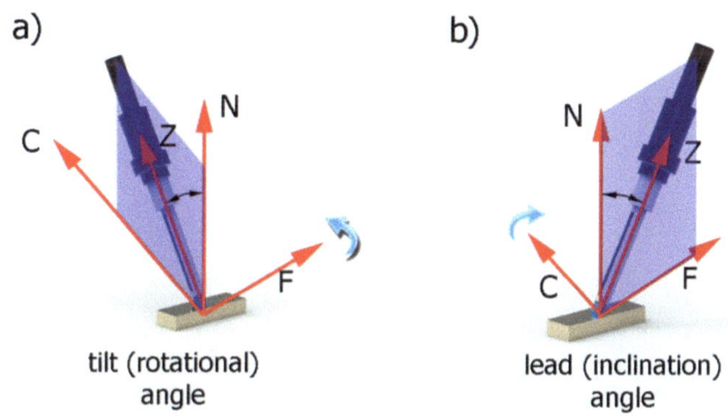

Fig. 5. Graphical presentation of: a) tilt angle, b) lead angle

To achieve the lead angle settings assumed in the experiment (0°, 15°, 30°, and 45°) on a 3-axis machine, the sample was oriented at the appropriate angle relative to the tool axis, as illustrated in Fig. 6.

Fig. 6. Graphical presentation setting the lead angle on a 3-axis machining center

3 Results

3.1 Surface Topography Analysis

Figure 7 illustrates the surface topography obtained after milling (7a) and after brushing with red fibers (7b) and blue fibers (7c) (without generated surface defects). After milling, characteristic marks corresponding to the imprint of the cutter edge on the surface are visible. After brushing, the imprint of the red and blue fibers can be observed on the surface.

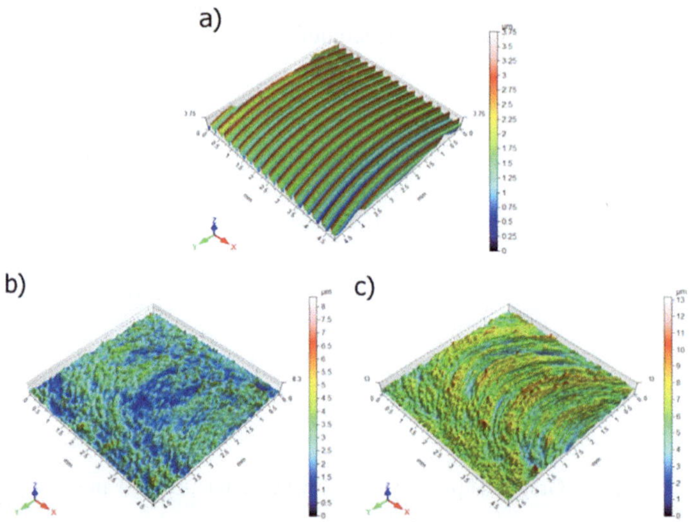

Fig. 7. Surface topography after: a) milling, b) brushing with red fibers, c) brushing with blue fibers

Since all samples were milled prior to brushing using identical cutting parameters, the average surface roughness after milling is indicated in Fig. 8 as a horizontal reference line with a value of Sa = 0.57 µm. The stiff blue fibers caused surface degradation (an increase in roughness), as shown in Fig. 8b. In contrast, brushing with the tool equipped with red fibers maintained (or slightly improved) the surface roughness obtained after milling (Fig. 8a).

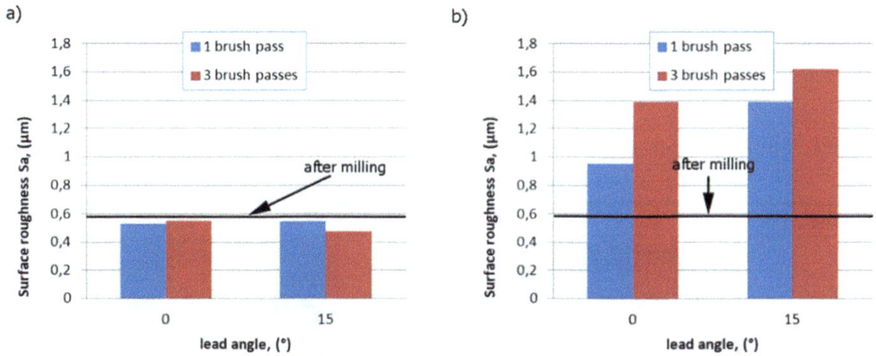

Fig. 8. Surface roughness after: a) brushing process with red fibers, b) brushing with blue fibers

3.2 Surface Defects Analysis

The average depth of the generated defects (before brushing) was 18 ± 5 µm (for the nominal setting of 20 µm) and 58 ± 9 µm (for the nominal setting of 60 µm). After the brushing process, the scratch depth was reduced or the defects were completely removed. The effect of brushing on the depth of surface defects using red and blue fibers is shown in Figs. 9 and 10, respectively.

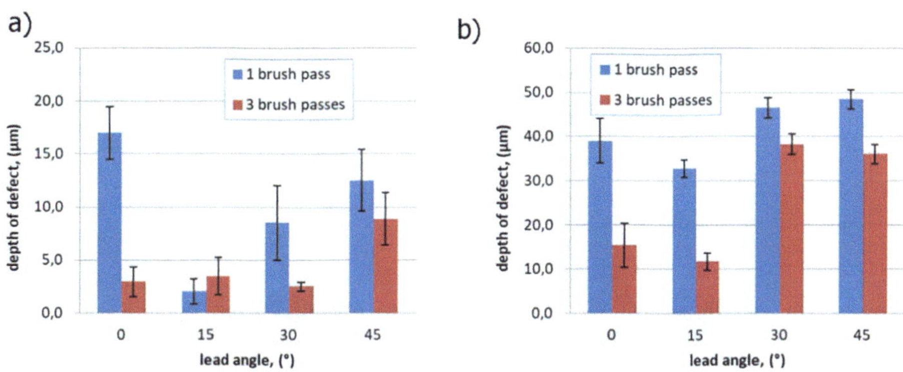

Fig. 9. Impact of brushing process on surface defect depth for red ceramic fibers at an initial defect depth of: a) 20 µm, b) 60 µm

For a lead angle of 0° (tool axis perpendicular to the machined surface), the effectiveness of defect removal is the lowest. This is because the brush's frontal surface rests on the workpiece, preventing the fibers from reaching the bottom of the generated defect.

For blue fibers, with a nominal scratch depth of 20 µm, using lead angles greater than zero resulted in complete removal of surface defects. In the case of a deeper scratch (nominal value 60 µm), complete defect removal was achieved at a lead angle of 15°.

The application of a 15° lead angle enabled the elimination of surface defects. This may be due to the large number of fibers simultaneously acting on the machined surface, compared to larger angles, where the fibers strike partially with their cylindrical surface, which is less effective. Introducing a positive lead angle allows for better penetration into the bottom of the surface defect.Table 1 illustrates the condition of surface defects after brushing with blue fibers for different lead angles.

Table 1. Surface defects after the brushing process with blue fibers for different lead angles

Lead angle (°)	1 pass	3 passes
0		
15		
30		
45		

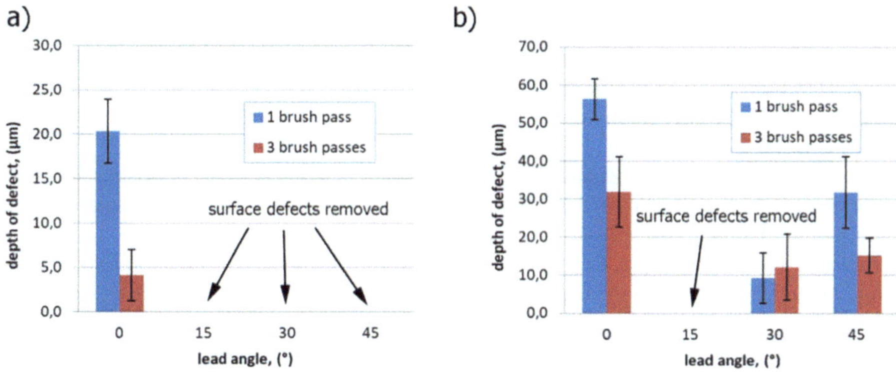

Fig. 10. Effect of brushing on the depth of surface defects for blue ceramic fibers at an initial defect depth of: a) 20 µm, b) 60 µm

4 Conclusions

This research analyzed how brushing conditions impact the effectiveness of removing surface defects.. Defects were generated at two depth levels, 20 and 60 µm. The influence of lead angle, type of brush fiber, and brushing tool pass count on surface defect removal in EN AW 2024 aluminum alloy components was investigated. The brushing process's effect on surface roughness was additionally analyzed. The experimental results are presented in the following conclusions:

- Ceramic brushes can be an effective tool for the removal of defects on the surface,
- A significant impact of the lead angle on the effectiveness of surface defect removal was observed,
- Brushing with red fibers maintained the surface roughness at the level obtained after milling, whereas the stiffer blue fibers caused an increase in surface roughness,
- Using blue fibers in the brushing tool results in higher effectiveness in removing surface defects,
- The highest effectiveness (reduction or complete removal of defects) was observed for a lead angle of 15°, due to better penetration into the bottom of the defect during brushing.

References

1. Liu, S., Wang, X., Liu, Z., Wang, Y., Chen, H.: Machined surface defects monitoring through VMD of acoustic emission signals. J. Manuf. Process. **79**, 587–599 (2022)
2. Jayakumar, N., Senthilkumar, G., Vignesh, K.S.: Investigation on surface defect ma-chining of AISI 4340 steel. Mater. Today Proc. **66**, 1189–1195 (2022)
3. Kujawińska, A., Diering, M., Rogalewicz, M., Żywicki, K., Hetman, Ł.: Soft modelling-based methodology of raw material waste estimation. In: Burduk, A., Mazurkiewicz, D. (eds.) Intelligent Systems in Production Engineering and Maintenance – ISPEM 2017, pp. 407–417. Springer International Publishing, Cham (2018). https://doi.org/10.1007/978-3-319-64465-3_39

4. Kujawińska, A., et al.: Methodology supporting the planning of machining allowances in the wood industry. In: Pérez García, H., Alfonso-Cendón, J., Sánchez González, L., Quintián, H., Corchado, E. (eds.) International Joint Conference SOCO'17-CISIS'17-ICEUTE'17, pp. 338–347. Springer International Publishing, Cham (2018). https://doi.org/10.1007/978-3-319-67180-2_33
5. Bao, L., Li, K., Zheng, J., Zhang, Y., Zhan, K., Yang, Z.: Surface characteristics and stress corrosion behavior of AA 7075-T6 aluminum alloys after different shot peening pro-cesses. Surf. Coat. Technol. **440**, 128481 (2022)
6. Kowalik, M., Trzepieciński, T., Kukiełka, L., Paszta, P., Maciąg, P., Legutko, S.: Experimental and numerical analysis of the depth of the strengthened layer on shafts resulting from roller burnishing with roller braking moment. Materials. **14**(19), 5844 (2021)
7. Grochała, D., Berczyński, S., Grządziel, Z.: Modeling of burnishing thermally toughened X42CrMo4 steel with a ceramic ZrO 2 ball. Arch. Civil Mech. Eng. **17**(4), 1011–1018 (2017)
8. Matuszak, J., Zaleski, K., Skoczylas, A., Ciecieląg, K., Kęcik, K.: Influence of semi-random and regular shot peening on selected surface layer properties of aluminum alloy. Materials. **14**(24), 7920 (2021)
9. Skoczylas, A., Zaleski, K., Matuszak, J.: Evaluation of the effectiveness of surface defect removal by slide burnishing. Appl. Sci. **15**(13), 7398 (2025)
10. Matuszak, J.: Comparative analysis of the effect of machining with wire and ceramic brushes on selected properties of the surface layer of EN AW-7075 Aluminium alloy. Adv. Sci. Technol. Res. J. **16**(2), 50–56 (2022)
11. Skoczylas, A., Zaleski, K.: Effect of centrifugal shot peening on the surface properties of laser-cut C45 steel parts. Materials. **12**(21) (2019)
12. Skoczylas, A., Kłonica, M.: Selected properties of the surface layer of C45 steel samples after slide burnishing. Materials. **16**(19), 6513 (2023)
13. Ciecieląg, K.: Effect of composite material fixing on hole accuracy and defects during drilling. Adv. Sci. Technol. Res. J. **15**(3), 54–65 (2021)
14. Ciecieląg, K., Skoczylas, A., Matuszak, J., Zaleski, K., Kęcik, K.: Defect detection and localization in polymer composites based on drilling force signal by recurrence analysis. Measurement. **186**, 110126 (2021)
15. Ciecieląg, K., Kęcik, K., Skoczylas, A., Matuszak, J., Korzec, I., Zaleski, R.: Non-destructive detection of real defects in polymer composites by ultrasonic testing and recurrence analysis. Materials. **15**(20), 7335 (2022)
16. Cariapa, V., Stango, R.J., Chen, L., Hermann, R.: Aspects of process model for automatic control of edge-deburring with filamentary brush. J. Eng. Ind. **114**(3), 294–300 (1992)
17. Chen, L., Stango, R.J., Cariapa, V.: A force-control model for edge-deburring with filamentary brush. J. Manuf. Sci. Eng. **123**(3), 528–532 (1999)
18. Kumar, S., Dornfeld, D.: Basic approach to a prediction system for Burr formation in face milling. J. Manuf. Process. **5**(2), 127–142 (2003)
19. Onstein, I.F., Bjerkeng, M., Martinsen, K.: Automated tool trajectory generation for Robotized deburring of cast parts based on 3D scans. Proc. CIRP. **118**, 507–512 (2023)
20. Matuszak, J., Kłonica, M., Zagórski, I.: Measurements of forces and selected surface layer properties of AW-7075 aluminum alloy used in the aviation industry after abrasive machining. Materials. **12**(22), 3707 (2019)
21. Stango, R.J., Cariapa, V., Zuzanski, M.: Contact zone force profile and machining performance of filamentary brush. J. Manuf. Sci. Eng. **127**(1), 217–226 (2005)
22. Shia, C.Y., Stango, R.J.: Analysis of a compliant honing tool for brushing cylindrical surfaces. J. Manuf. Sci. Eng. **119**(3), 441–444 (1997)

Parameterization of the Grinding Process with Abrasive Belts for Complex Surfaces of Surgical Instruments

Jędrzej Bożek[1](✉), Adam Poschwald[1], and Adam Hamrol[2]

[1] Aesculap Chifa, Nowy Tomysl, Poland
jedrzej.bozek@bbraun.com
[2] Faculty of Mechanical Engineering, Poznan University of Technology, Poznan, Poland

Abstract. The article presents research on optimizing grinding parameters for abrasive belt grinding of surgical tools in the context of replacing manual processes with automated machining. The main objective was to determine the minimum acceptable surface quality after grinding that would enable achieving the required quality after vibratory finishing. The research was conducted on samples replicating the shape of plaster cutting scissors, using belts with grits P50, P60, and P80 in various grinding strategies. The influence of the number of passes (1, 2, 3) and feed rate (F100, F200, F300) on surface roughness (*Ra*, *Rz*) before and after vibratory finishing was analyzed. The conducted ANOVA test for trials 1–9 showed that at a significance level of $\alpha = 0.05$, for a given longitudinal feed rate, surface roughness does not depend on either belt grit size or the number of passes. This result may stem from the specific characteristics of surgical tool materials and the use of machining parameters atypical for coarse belts. The findings indicate that a grinding strategy with multiple passes at higher feed rates (F200, F300) is less favorable than a strategy with a single pass but lower longitudinal feed rate.

Keywords: Abrasive Belt Grinding · Surgical Tools · Vibratory Finishing

1 Introduction

In the manufacturing processes of surgical instruments, grinding and vibratory finishing play a significant role. They ensure the appropriate surface structure of the instruments, which directly affects their functionality, ease of sterilization, and operational durability [1]. Despite considerable progress in the automation of surgical instrument production, due to their complex geometry, small dimensions, and high requirements for precision and quality, grinding operations are largely performed manually using abrasive belts [1, 2].

M. Hryb et al. (Eds.): MANUFACTURING 2026, LNME, pp. 147–156, 2026.
https://doi.org/10.1007/978-3-032-23038-6_11

Manual belt grinding (Fig. 1) allows for efficient removal of excessive material and enables an experienced operator to compensate for shape deviations resulting from technological operations preceding the grinding process [2]. However, the inability to parameterize grinding conditions means that operators have different preferences when selecting the type and characteristics of the abrasive belt, the applied pressure, and the machining trajectory. As a result, the grinding process is characterized by low repeatability and a high degree of dependence on the operator's competence [1].

Fig. 1. Manual grinding in Aesculap Chifa Poland

Ergonomic aspects of the grinder's work should also be considered. Operators of belt grinding machines are exposed to occupational risk factors, including contact with moving machine components, noise levels of 85–95 dB, abrasive dust, forced body posture, and task monotony [3].

Considering these factors, at Aesculap Chifa Polska (ACP), one of the world's leading manufacturers of surgical instruments, work has been undertaken to gradually replace manual grinding with automated grinding performed on grinding machines. However, since the strategies used by manual grinders are not parameterized—they are part of so-called tacit knowledge [1]—before introducing automated grinding, it is necessary to conduct research to determine grinding parameters that ensure effective and repeatable compliance with the quality requirements imposed on surgical instruments. These requirements relate to dimensions associated with dimensional and shape accuracy, as well as surface structure and the condition of the surface layer. At the same time, process parameters must ensure high machining efficiency.

The aim of the research described in this study was to obtain information on the influence of grinding process efficiency and abrasive belt grit size on the structure of grinded surfaces, defined by roughness parameters. An additional objective was to indicate the extent to which surface roughness after grinding is improved by vibratory finishing.

In the study, two surface roughness parameters of the grinded surface were determined: *Ra* and *Rz*. Roughness *Ra* is the quality standard applied at ACP and is also one of the most used linear surface roughness parameters [4, 5]. However, the *Ra* parameter has significant limitations. As research indicates [4–6], *Ra* takes little account of individual sharp peaks or depressions on the surface. This means that by averaging all surface deviations, *Ra* can mask extreme roughness values that may be crucial for the functionality of a medical product, particularly in terms of sealing, friction, or biocompatibility.

For the reasons mentioned above, the *Rz* parameter was also measured in the study. *Rz* is a two-dimensional parameter calculated based on five individual measurement segments [6]. Unlike *Ra, Rz* is sensitive to scratches, damage, and surface contamination [5, 6], which allows the detection of extreme height deviations that may be overlooked by *Ra.* By considering differences between the highest peaks and the deepest valleys, *Rz* provides complementary information about the actual surface topography, which is particularly important in the context of medical instrument production.

In most studies, the relationship between the roughness of the grinded surface and the grit size of the abrasive belt is linear—the finer the grains, the lower the roughness of the grinded surface. Li et al. [7], in research on belt grinding of TC4 titanium alloy, demonstrated a strong linear correlation between process parameters and surface roughness ($R^2 = 0.9547$), with roughness being most sensitive to changes in belt grit size. Similar observations were confirmed by Wang et al. [8] for bearing steel grinding, where smaller abrasive belt grit size led to a reduction in both the height of micro-irregularities and the roughness of the grinded surface.

However, not all studies confirm this simple linear relationship. Mezghani S et al. [9], in research on finishing belt grinding, demonstrated the existence of an optimal grit size for achieving the lowest surface roughness. In their experiments, conducted with grain sizes ranging from 9 to 80 μm, grains of 30 μm provided better roughness reduction than finer grains of 9 μm, indicating a non-linear relationship and the existence of an optimum grit size for specific machining conditions.

2 Experimental Procedure

The cognitive objective of the conducted research was to determine the extent to which the relationships described in the literature [10–15] apply to the specific conditions of belt grinding surgical instruments at ACP.

The practical objective of the research was to determine the influence of abrasive belt grain size and process performance parameters (longitudinal feed, number of passes) on the surface roughness of shaped samples made of hardened X20Cr13 steel, commonly used to produce surgical instruments, and to establish the degree of surface roughness reduction after grinding during vibratory finishing.

The research was carried out on the experimental grinding setup shown in Fig. 2. The machine was equipped with a dual track grinding unit with integrated electric drives and four electromechanical axes using servo technology with energy recovery. The machine was also equipped with a CNC system with software controlling the grinding process and a synergistic sensor system for process monitoring.

Figure 3 shows the kinematic diagram of the process.

The diagram indicates: the longitudinal feed of the grinded surface relative to the abrasive belt (v_f), the infeed of the grinded surface in the direction perpendicular to the belt per pass (Δf_n), and the linear speed of the abrasive belt (v_s).

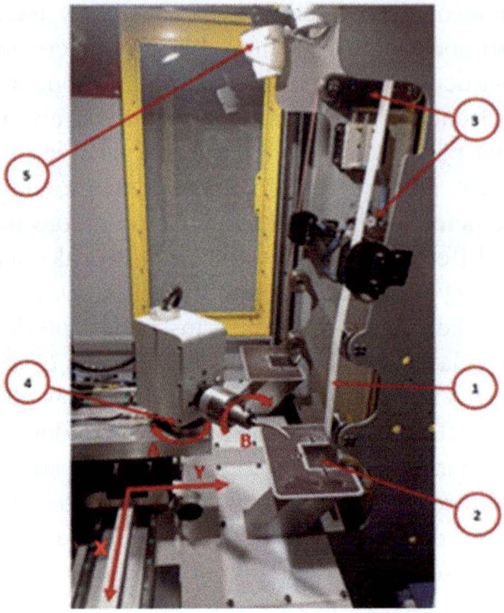

Fig. 2. Experimental grinding setup (1 – abrasive belt, 2 – dust extraction, 3 – belt tensioning system, 4 – clamping system, 5 – vision system)

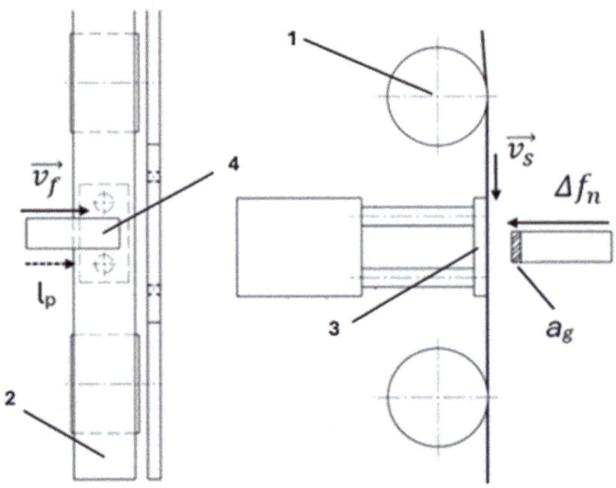

Fig. 3. Kinematic Diagram of the experimental grinding setup (1 – support rollers, 2 – abrasive belt, 3 – pneumatic support, 4 – workpiece)

The number of longitudinal passes of the belt (lp) and the infeed per pass (Δf_n) were selected so that in all tests the same excessive material value was removed ($a_g = 0.5$ mm). This means that grinding efficiency, measured by the operation duration, was determined by the longitudinal feed of the abrasive belt (v_f) and the number of longitudinal passes of the belt (lp).

An additional process evaluation metric was the grinding efficiency index E_s, defined by the formula:

$$E_s = \frac{a_g}{\sum \Delta f_n} \tag{1}$$

where: a_g – the amount of excessive material removed,

$\sum \Delta f_n$ – the total infeed of the grinded surface of the sample toward the belt.

The higher the E_s index value, the greater the portion of grinding energy that is "converted" in the process into effective material removal, and consequently, the smaller the amount lost to inefficient friction of the abrasive belt against the grinded surface or its plastic deformation. As a result, during the grinding process — due to inefficient cutting, less heat is generated, and the grinded surface is less exposed to thermal damage.

After grinding, the samples were subjected to vibratory finishing [16] in the device shown in Fig. 4. Standard process parameters were selected for vibratory finishing, as used in the processing of standard medical tools manufactured at ACP.

Fig. 4. Vibratory polishing machine and the abrasive media used in the process

The research plan is presented in Table 1.

Table 1. Research plan

<table>
<tr><td colspan="5">Grinding</td><td>Vibratory Polishing</td></tr>
<tr><td rowspan="3">Trial number</td><td rowspan="3">Number of longitudinal passes [lp]</td><td colspan="3">Longitudinal feed V_f [mm/min]</td><td rowspan="3">Power 3,5 kW
Supply voltage 400 V
Frequency 50 Hz
Processing time 6 h</td></tr>
<tr><td colspan="3">Grinding belt grit</td></tr>
<tr><td>P50</td><td>P60</td><td>P80</td></tr>
</table>

(continued)

Table 1. *(continued)*

Grinding					Vibratory Polishing
1–3	1	F100	F100	F100	
4–6	2	F100	F100	F100	
7–9	3	F100	F100	F100	
10–12	2	F200	F200	F200	
13–15	3	F300	F300	F300	

P50, P60, P80 - grit sizes – To maintain the repeatability of the experiment, a new, fresh abrasive belt was used for each test series.
F (100;200;300) - Longitudinal feed $V_f = 100, 200, 300$ mm/min
The sample's pressure force was not directly measured during the experiment; however, it can be approximated based on the $\sum \Delta f_n$ value, as the relationship between infeed and applied force is proportional.
Fixed parameters:
Grinding belt 3 M Cubitron 784F, backing weight YF
Grinding speed $v_s = 30$ m/s
Excessive material a_g - 0,5 mm

3 Results

Figure 5 presents the results of *Ra* and *Rz* roughness measurements obtained in individual grinding trials and after vibratory finishing.

The results indicate no monotonic relationship between belt grit size and surface roughness (i.e., the larger the grains, the higher the roughness). After grinding with feed rate F100 and a single pass using a P50 grit belt, a lower surface roughness was achieved compared to grinding with P60 and P80 grit belts. Similar relationships were observed for grinding with two and three passes. Non-monotonic dependencies between grain size and surface roughness were also obtained for grinding with feed rate F200 and two passes, as well as with feed rate F300 and three passes. Thus, the results are consistent with those presented in the publication, confirming the existence of an optimal surface grit size [9] and deviating from known predictive models of surface roughness in grinding [17–19]. These deviations may result from the specific properties of surgical tool material and the use of parameters atypical for coarse belts.

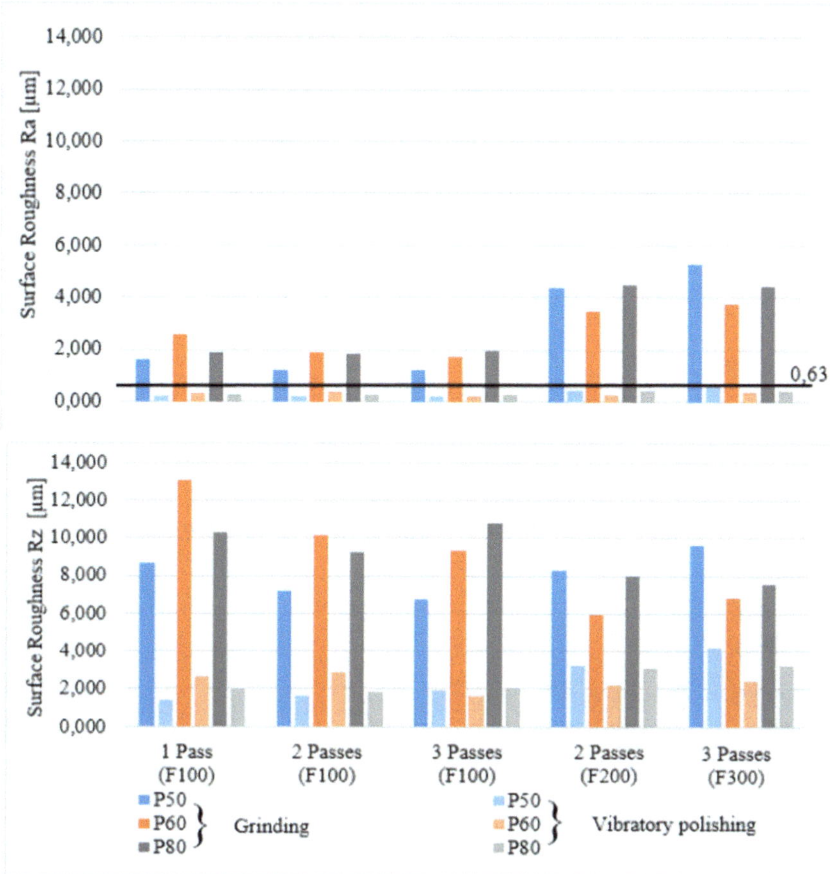

Fig. 5. Comparison of surface roughness *Ra* and *Rz* after grinding and vibratory finishing

From the charts in Fig. 5, it follows that, assuming the maximum allowable surface roughness after vibratory finishing, $Ra < 0.63$, this requirement is met in all conducted trials. The improvement in the *Rz* parameter is significantly smaller than that of *Ra* (*Ra* roughness decreases on average by 5–8 times, *Rz* by 2–3 times), as shown in Figs. 6a and 6b. This can be explained by the fact that vibratory finishing exhibits a limited depth of influence on surface irregularities after grinding. When determining the *Rz* parameter, extreme surface depressions are considered, which results in a smaller relative change in this value.

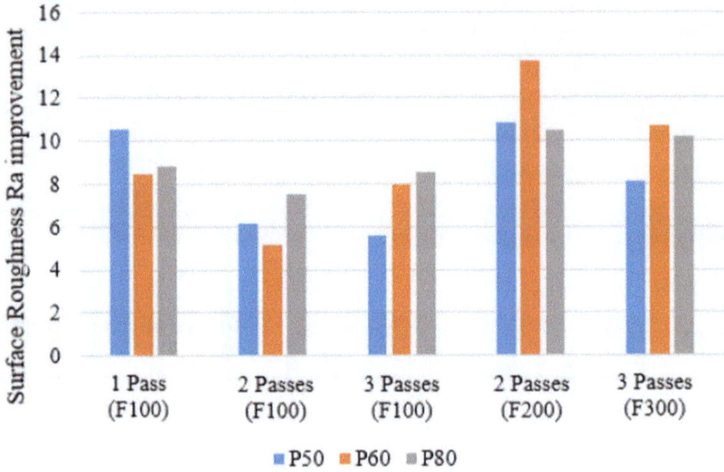

a Improvement of surface roughness *Ra* after vibratory finishing

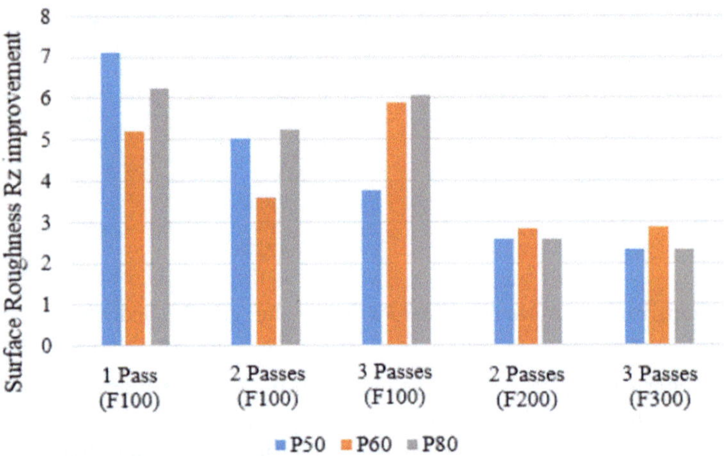

b Improvement of surface roughness *Rz* after vibratory finishing

Fig. 6. a. Improvement of surface roughness *Ra* after vibratory finishing. b. Improvement of surface roughness *Rz* after vibratory finishing

Considering the non-monotonic relationship between the roughness of the grinded surface and the size of abrasive grains, as well as the fact that, regardless of the grinding parameters used in the study, the surface roughness after vibratory finishing meets technological requirements, it can be concluded that the decisive criterion for selecting grinding parameters is efficiency and grinding time. Figure 7 presents the values of the grinding efficiency coefficient E_s, and the grinding times obtained in individual trials.

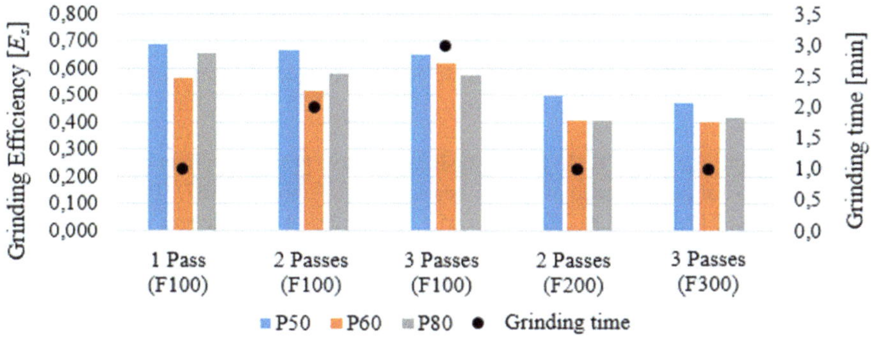

Fig. 7. Grinding Efficiency E_S

The highest grinding efficiency index E_s, as well as the shortest grinding time, were obtained when grinding with a P50 grit belt, using one pass and a feed rate of F100. Conversely, the lowest E_s value was achieved—as expected—when grinding with two or three passes using longitudinal feed rates of F200 and F300. This means that a grinding strategy with multiple passes and higher feed rates (F200, F300) is less advantageous than a strategy with a single pass and a smaller longitudinal feed rate.

4 Conclusions

The research presented in this article is part of the process of replacing manual grinding with abrasive belts on contoured surfaces of surgical instruments with automated grinding at ACP. It serves as a starting point for further work on parameterizing and optimizing the grinding process. The obtained results indicate that the process exhibits specific characteristics, making it unjustified to apply relationships known from the literature or from other applications of abrasive belt grinding. The dependencies presented in this article must be confirmed by studies covering a broader range of grinding parameters, the condition of the surface layer, and process variables such as grinding force and the temperature of the grinded surface.

References

1. Hamrol, A., Hoffmann, M., Lisek, M., Bożek, J.: The quality of surgical instrument surfaces machined with robotic belt grinding. Materials. **16**, 630 (2023)
2. Seavey, R.: High-level disinfection, sterilization, and antisepsis: current issues in reprocessing medical and surgical instruments. Am. J. Infect. Control. **41**, S111–S117 (2013)
3. Hoffmann Group: Ochrona przed hałasem w nowoczesnym środowisku pracy. https://www.hoffmann-group.com/PL/pl/perp/wiedza-techniczna/przewodniki-online/przewodnik-po-technice-szlifowania-i-ciecia/halas/e/513903/. Accessed 12 July 2025
4. Whitehead, K.A., Verran, J.: The effect of surface topography on the retention of microorganisms. Food Bioprod. Process. **84**, 253–259 (2006)
5. Volk, R.: Surface texture metrology. In: Hügel, H., Graf, T. (eds.) Laser in der Fertigung, pp. 1–20. Springer Vieweg, Berlin, Heidelberg (2018)

6. Mínguez-Martínez, A., et al.: Results of a surface roughness comparison between stylus instruments and confocal microscopes. Materials. **15**, 5495 (2022)
7. Li, Y., Zhang, C., Wang, Y., Guo, J., Xu, J.: Sensitivity study of surface roughness process parameters in belt grinding titanium alloys. Meta. **13**, 1825 (2023). https://doi.org/10.3390/met13111825
8. Wang, Y., Hou, B., Wang, F., Ji, S.: Analysis of abrasive grain size effect of abrasive belt on material removal performance of GCr15 bearing steel. Tribol. Int. **167**, 107380 (2022)
9. Mezghani, S., El Mansori, M., Zahouani, H.: New criterion of grain size choice for optimal surface texture and tolerance in belt finishing production. Wear. **266**(5–6), 578–580 (2009)
10. Pandiyan, V., Caesarendra, W., Tjahjowidodo, T., Praveen, G.: Predictive modelling and analysis of process parameters on material removal characteristics in abrasive belt grinding process. Appl. Sci. **7**, 363 (2017)
11. Zhu, D., et al.: Robotic grinding of complex components: a step towards efficient and intelligent machining—challenges, solutions, and applications. Robot. Comput. Integr. Manuf. **65**, 101956 (2020)
12. Wang, J., Xu, J., Zhang, X., Ren, X., Song, X., Chen, X.: An investigation of surface corrosion behavior of Inconel 718 after robotic belt grinding. Materials. **11**, 2440 (2018)
13. Syreyshchikova, N.V., Pimenov, D.Y., Gupta, M.K., Nadolny, K., Giasin, K., Sharma, S.: Establishing the relationship between cutting speed and output parameters in belt grinding on steels, aluminum and nickel alloys: development of recommendations. Materials. **14**, 1974 (2021)
14. Huang, Y., Fan, W., Guo, B., Chen, W.: Roughness prediction and performance analysis of data-driven superalloy belt grinding. Front. Mater. **9**, 765401 (2022)
15. Neşeli, S., Yaldız, S., Türkeş, E.: Determining the optimum process parameter for grinding operations using robust process. J. Mech. Sci. Technol. **26**, 3587–3595 (2012)
16. Mass Finishing Inc: Vibratory finishing process fundamentals and applications in metal surface treatment. J. Manuf. Process. **15**, 234–248 (2020)
17. Fan, W., Wu, C., Wu, Z., Liu, Y.: Prediction of surface roughness of abrasive belt grinding of superalloy material based on RLSOM-RBF. Materials. **14**, 5701 (2021)
18. Pandiyan, V., Caesarendra, W., Tjahjowidodo, T., Praveen, G.: Predictive modelling and analysis of process parameters on material removal characteristics in abrasive belt grinding process. Appl. Sci. **7**(4), 363 (2017)
19. Jourani, A., Dursapt, M., Hamdi, H., Rech, J., Zahouani, H.: Effect of the belt grinding on the surface texture: modeling of the contact and abrasive wear. Wear. **259**(7–12), 1137–1143 (2005)

Comparative Erosion Analysis of Ti6Al4V and AISI 304 L Under Different Excitation Frequencies Using PWJ

Gabriel Stolárik[1](✉), Alice Chlupová[2], Jakub Poloprudský[2], Akash Nag[1], and Sergej Hloch[1]

[1] Faculty of Manufacturing Technologies, Technical University of Košice, With a Seat in Prešov, Prešov, Slovakia
gabriel.stolarik@tuke.sk
[2] Institute of Physics of Materials, Czech Academy of Sciences, Brno, Czechia

Abstract. The study focuses on the controlled modification of the surface topography of Ti6Al4V alloy and AISI 304 L stainless steel using pulsating water jet (PWJ) technology. While previous research analyzed these materials separately, this work compares, for the first time, the characteristics of surfaces generated on both materials under identical conditions. Surface modification was carried out using linear and perpendicularly interlaced trajectories at frequencies of $f_s = 20$ kHz and $f_s = 40$ kHz. At $f_s = 20$ kHz, a higher degree of erosion was observed on Ti6Al4V. Conversely, at $f_s = 40$ kHz, the AISI 304 L steel exhibited significantly higher surface roughness S_a, reaching a maximum of $S_a = 25.56$ μm for the linear trajectory, nearly three times the value (9.17 μm) measured on Ti6Al4V under the same conditions. This phenomenon results from the distinct responses of the materials to cyclic loading, governed by their specific mechanical properties, particularly the ratio of strength, toughness, and plasticity, where Ti alloys are subject to continuous erosion and AISI steels require exceeding a critical load threshold, after which rapid material removal occurs. The results indicate that the material's ability to absorb impact energy through plastic deformation plays a key role in the formation of surface topography. At the same time, the perpendicular interlaced trajectory can machine the surface more homogeneously and isotropically, which shows its advantage over the linear trajectory. The study, therefore, contributes to a deeper understanding of PWJ interactions with metallic surfaces and provides data to enable the targeted creation of surface structures with defined topographical characteristics.

Keywords: Surface Modification · Surface Roughness · Pulsating Water Jet · Droplet Erosion · Frequency

1 Introduction

Surface modification using pulsating water jet (PWJ) technology represents a promising approach for controlled alteration of surface topography and functional properties of materials. Unlike commonly used techniques such as plasma spraying or grit blasting

M. Hryb et al. (Eds.): MANUFACTURING 2026, LNME, pp. 157–171, 2026.
https://doi.org/10.1007/978-3-032-23038-6_12

[1], PWJ enables roughening processes using pure water without the addition of abrasive media, thereby eliminating the undesirable effects of the heat-affected zone [2] and surface contamination [3]. The principle of the method lies in the action of discrete water segments impacting the material surface, generating impact pressure known as "water hammer pressure." Repeated impacts of water segments generate localized cyclic loading and cause material fatigue, resulting in more efficient material removal compared to a continuous water jet (CWJ) [4].

The intensity of induced erosion is determined by a set of hydraulic and kinematic parameters of the jet, whereas the mechanical properties of the target material govern its erosion resistance. The erosion process can be observed in successive stages corresponding to the gradual increase in the number of impacts over time [5]. PWJ technology offers extensive control over the distribution of water clusters and, consequently, precise temporal loading of the material, enabling regulation of the erosion process [6]. The fundamental means of control include adjusting the frequency of generated water segments and using linear or perpendicularly interlaced trajectories with specific traverse speeds [7]. By tuning these parameters, it is possible to regulate material loading and initiate the desired erosion stage precisely. Incubation phases can, for example, be used to modify weld joints, transforming tensile stresses into compressive ones, thereby improving the joint's mechanical properties. This phenomenon was also confirmed in a study [8], in which the authors investigated the extent of changes in mechanical properties on TIG-welded joints made of AISI 304 steel. They found that, when pressures of 20, 40, and 60 MPa were applied, an increase in compressive residual stress occurred (reaching a maximum of 535.46 MPa at a pressure of 20 MPa and a traverse speed of 2 mm·s^{-1}) compared to the base material (240.98 MPa). The magnitude of this stress was directly dependent on the applied traverse speed in the range of 0.25–5.5 mm·s^{-1}, i.e., on the PWJ exposure time on the samples.

Conversely, advanced erosion phases, characterized by material removal, can be applied to surface roughening and enhance adhesive properties. Such modified surfaces with optimized roughness parameters are used in implantology, where surface topography significantly influences osseointegration and long-term implant stability [9]. The current relevance of this topic has led to various experimental studies focused on the application of PWJ technology for controlled roughening and the characterization of advanced erosion stages to verify its qualification and application potential. For example, in the study [10], the authors confirmed that at a pressure of 70 MPa and a nozzle diameter of 0.8 mm, the surface topography of AISI 304 L steel can be modified using a frequency of 40 kHz. They found that a perpendicular interlaced trajectory enables higher surface isotropy (maximum of 96.74% at $v = 10$ mm·s^{-1}) compared to a linear trajectory (maximum of 88.23% at $v = 5$ mm·s^{-1}). This indicates a more uniform spatial distribution of roughness, resulting in a more homogeneous surface. The same technological parameters of pressure and nozzle diameter were also applied under submerged hydrostatic conditions with a water column height of 150 mm [11]. The results demonstrated a substantial reduction in the PWJ effect (by up to 97%), attributed to the absorption of the high-velocity water jet's kinetic energy by the surrounding static water. In another study focused on titanium alloy [12], the authors demonstrated the ability to modify the surface of the Ti6Al4V alloy at frequencies $f_s = 20$ and 40 kHz, using the same pressure and nozzle diameter as in the previous studies. The results showed a significant decrease in surface roughness at 40 kHz compared to 20 kHz, which was attributed to the breakup of the water jet into smaller-mass clusters with lower kinetic energy. At the same time, however, an increase in subsurface microhardness by 25.3% compared to the base material was observed at 40 kHz. For this reason, the results do not allow an unambiguous determination of which frequency is more suitable for the surface modification process. Finally, a process using a higher operating pressure of 100 MPa was also proposed [13], which could reduce the energy demand of the modification process and increase the potential competitiveness of the PWJ method [14].

In previously published studies, the influence of PWJ process parameters (e.g., traverse speeds of $v = 1–5$ mm·s^{-1} for linear and $v = 2–10$ mm·s^{-1} for perpendicular trajectories, and excitation frequencies of $f_s = 20$ and 40 kHz) was analyzed individually on single materials, most commonly Ti6Al4V [12, 14] or AISI 304 L [10, 11]. These materials were selected primarily because both exhibit excellent corrosion resistance, which is important for maintaining surface integrity after the PWJ process. At the same time, the Ti6Al4V alloy is widely used in biomedical applications, while AISI 304 L steel exhibits different plastic properties, thus enabling a comparison of the development of erosion mechanisms between materials with different deformation behavior. In existing studies, erosion has mostly been analyzed in terms of trajectory and traverse speed variations for a single material. As a result, these findings provide only a partial view of the erosion response of individual materials without direct cross-material comparison. For a comprehensive understanding of the evolution of erosion processes, it is necessary to compare the effects of excitation frequencies of 20 and 40 kHz across different materials, which has not yet been systematically investigated.

The primary objective of this work is therefore to experimentally investigate the erosion characteristics of selected materials and to compare their responses under strictly controlled, identical loading conditions (i.e., identical distribution of water segments). The main contribution of this study lies in identifying and quantifying a potential frequency-dependent inversion in the relative erosion response between the materials, which has not yet been systematically explored.

2 Material and Methods

2.1 Material

In this study, titanium alloy Ti6Al4V and stainless steel 304 L were used as experimental materials. The selection of these materials was based on their excellent corrosion resistance, which is essential for maintaining surface integrity during and after experimental testing. Titanium alloys are characterized by low density and high strength and fracture toughness, making them suitable for aerospace and biomedical applications. On the other hand, stainless steel AISI 304 L has lower strength than titanium alloys; however, it offers superior plasticity. A comparative evaluation of the erosion process on these two material types enables a deeper understanding of the erosion-damage mechanism and the development of surface topography as a function of their different mechanical properties and structural behavior. Table 1 presents a comparison of the mechanical properties of both analyzed materials.

Table 1 Comparison of the mechanical properties of Ti6Al4V and AISI 304 L [10, 12]

	Tensile strength R_m [MPa]	Yield strength $R_{p\,0.2}$ [MPa]	Elongation A [%]	Hardness H [HV]
Ti6Al4V	950	880	14	349
AISI304L	721	239	43	210

2.2 Experimental Method

The experiments in comparative studies were carried out using hybrid technology based on an ultrasonically excited pulsating water jet (PWJ). This technology is characterized by an acoustic assembly that induces pressure fluctuations into a high-pressure water stream entering the acoustic chamber. These pressure oscillations, after passing through the nozzle, are transformed into a high-velocity jet with fluctuating axial velocity. This process leads to the gradual breakup of the continuous water stream into discrete water segments that impact the material surface at a specific frequency. In these experiments, a supply pressure of $p = 70$ MPa, a nozzle diameter of $\mathrm{d} = 0.8$ mm, and two water cluster generation frequencies, $f_s = 20$ and 40 kHz, were used. In practice, this corresponds to a different number of water clusters impacting on the target surface per unit time (Table 2). For each experiment, the technology was set to the optimal parameters of the acoustic chamber length "l_c" according to [15] and the standoff distance "z" according to [16], ensuring the utilization of the maximum erosion potential of the generated water clusters. Moreover, at a constant operating pressure "p", the water clusters generated at 40 kHz have a smaller volume compared to those generated at 20 kHz. This factor contributes to variations in the magnitude of impact pressure and, consequently, in the jet's erosion potential [17]. The experiments were conducted using both linear and perpendicularly interlaced trajectories, with traverse speeds of $v = 1$–5 mm·s^{-1} for the linear mode and doubled speeds of $v = 2$–10 mm·s^{-1} for the perpendicular interlaced mode (Fig. 1). The doubled traverse speed in the perpendicular interlaced mode was selected to maintain an equal number of water cluster impacts per unit surface area, thereby ensuring an objective comparison of the two trajectory types without altering the technological parameter that determines the number of impacts.

Table 2. Technological parameters of experiments.

p [MPa]	d [mm]	f_s [kHz]	l_c [mm]	z [mm]	Trajectory	v [mm·s^{-1}]	Material
70	0.8	20	24	43	Linear	1, 2, 3, 4, 5	Ti6Al4V AISI304L
					Perpendicular interlaced	2, 4, 6, 8, 10	
		40	12	77	Linear	1, 2, 3, 4,	
					Perpendicular interlaced	2, 4, 6, 8, 10	

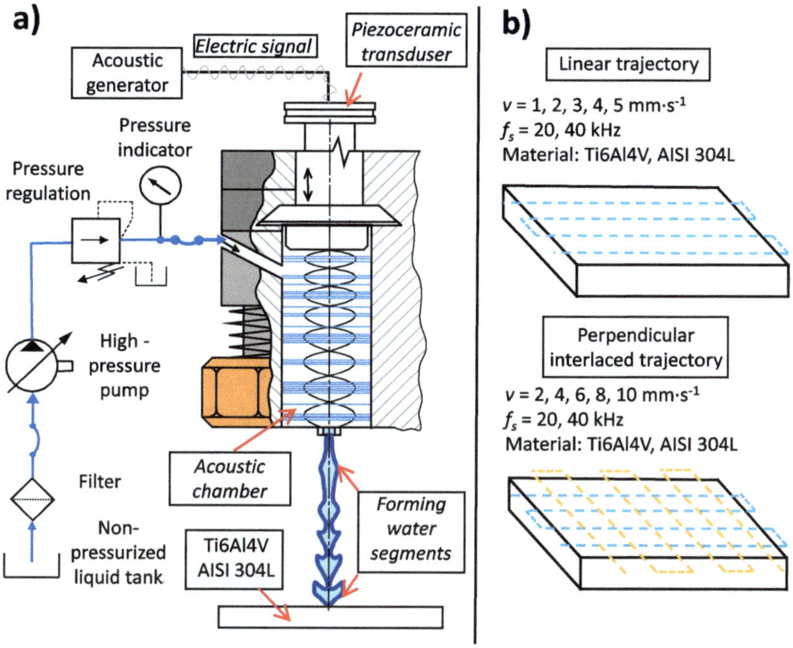

Fig. 1. a) experimental setup of PWJ technology, b) trajectories and parameters of the modification process

2.3 Measurement Method

Measurements were performed on five 3 × 3 mm areas of each treated sample. This area was chosen in accordance with the ISO 25178-3 standard, where it was necessary to obtain an area larger than the L-filter size of 2.5 mm to adequately obtain roughness parameters. The S_a parameter represents the arithmetic mean height of the surface within the measured area and provides quantitative information about the level of surface irregularities generated during treatment. In addition, a set of height parameters (S_p and S_v) and hybrid parameters (S_{sk} and S_{ku}) were analyzed to enable a more precise qualification of the surface topography. These supplementary characteristics allow the identification of valley–peak asymmetry (S_{sk}) and the prominence of extreme peaks or depressions (S_{ku}). The probability of functional properties was also supplemented with the isotropy parameter and S_{tr}. Surface characterization was performed using a MicroProf FRT optical profilometer that measures on the principle of spectral focus variation (FV). This device analyzes reflected white light using a spectrometer and a calibration table, enabling precise determination of surface height irregularities. The obtained 3D surface scans were subsequently processed in Mountains SPIP 10 software after applying S-L filters. The resulting roughness values were expressed as arithmetic means from five measurements, along with standard deviations, and visualized in line charts.

3 Results and Discussion

The figure shows 3D reconstructions of surfaces after the PWJ treatment process with a frequency of $f_s = 20$ kHz. It is evident that there is a difference in erosion topology between linear ($v = 1$ mm·s^{-1}) and perpendicular interlaced ($v = 2$ mm·s^{-1}) trajectories on the surfaces of Ti alloy and AISI steel. This difference in erosion topology also affects the overall roughness parameters achieved, which are discussed in the following chapters (Fig. 2).

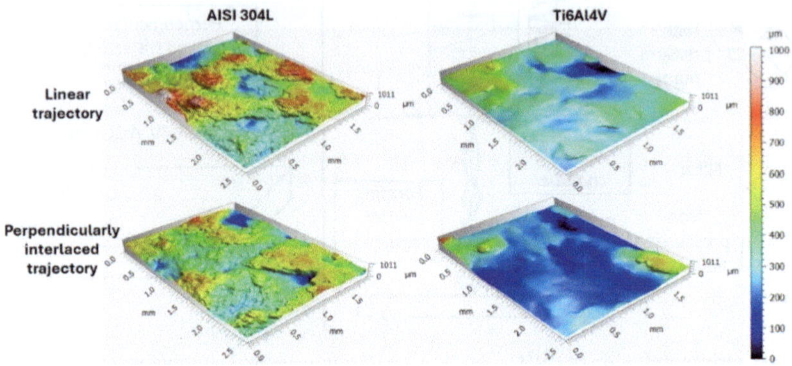

Fig. 2. 3D reconstruction of the surface after PWJ process with a pressure of 70 MPa, a nozzle diameter of 0.8 mm, and a linear ($v = 1$ mm·s^{-1}) and perpendicular interlaced ($v = 2$ mm·s^{-1}) trajectory for AISI 304 L and Ti6Al4V.

3.1 Modified Surfaces of Ti6Al4V and AISI 304 L at $f_s = 20$ kHz

Table 3 presents the average measured values of the surface roughness parameter S_a together with the calculated standard deviation for the surfaces of Ti6Al4V and AISI 304 L materials. The results clearly show that the frequency $f_s = 20$ kHz caused a pronounced increase in surface roughness for both applied trajectories, while differences in roughness values can also be observed between the two material types.

Table 3. Average measured roughness of S_a on Ti6Al4V and AISI 304 L at $f_s = 20$ kHz

	Linear trajectory					Perpendicularly interlaced trajectory				
v [mm·s^{-1}]	1	2	3	4	5	2	4	6	8	10
	Ti6Al4V									
S_a [µm]	67.92 ±6.71	61.41 ±2.28	37.32 ±9.92	44.37 ±1.45	24.78 ±0.61	83.86 ±4.20	39.40 ±4.17	17.87 ±0.72	6.40 ±1.27	3.14 ±0.23
	AISI 304 L									
S_a [µm]	159.23 ±1.07	161.53 ±10.47	2.71 ±0.19	1.38 ±0.04	1.37 ±0.11	185.83 ±17.15	64.27 ±1.06	15.93 ±1.78	2.79 ±0.06	1.24 ±0.02

Qualitative evaluation using areal surface roughness parameters (S_a) showed that, as traverse speed increases, the erosion potential of the target material decreases. This effect is attributed to a reduction in exposure time resulting from fewer water cluster impacts per unit area. This observation is consistent with the findings of [18], where increasing the traverse speed from $v = 200$ to $1000\ \text{mm}\cdot\text{s}^{-1}$ led to a decrease in the surface roughness parameter R_a, attributed to the reduced effective material loading time. A similar trend was reported in [19], where varying the traverse speed ($v = 200$–$1000\ \text{mm}\cdot\text{s}^{-1}$) together with the stand-off distance ($z = 25.4$–44.4 mm) significantly influenced both material removal and erosion mechanisms. Such erosion development behavior is also comparable with the study [20], confirming that it is typical for the PWJ modification mechanism. The extent of the resulting surface damage depends on the mechanical properties of the target material, as defined in [19]. This was also confirmed in the present comparison between Ti6Al4V and AISI 304 L (Fig. 3). At a frequency of 20 kHz, the most pronounced erosion on linear trajectories was observed at traverse speeds of $v = 1$ and $2\ \text{mm}\cdot\text{s}^{-1}$. The average roughness measured at v = 1 and $2\ \text{mm}\cdot\text{s}^{-1}$ was $S_a = 159.23 \pm 1.07$ μm and 161.53 ± 10.47 μm, respectively, on AISI 304 L surfaces, which are more than twice the values measured on Ti6Al4V ($S_a = 67.92 \pm 6.71$ μm and 61.41 ± 2.28 μm). This is attributed to the lower hardness of AISI 304 L compared with Ti6Al4V. Although AISI 304 L exhibits superior plasticity, localized cyclic loading exceeded its recovery capacity, resulting in more pronounced erosion craters. Conversely, the higher hardness of Ti6Al4V improves its erosion resistance, minimizing the formation of surface irregularities and ensuring greater material stability under identical loading conditions. This was also confirmed in [19], where modified CP-Ti showed higher surface roughness than Ti6Al4V [20]. However, the results of studies [10, 13] indicated a negative skewness parameter S_{sk} (maximum -2.61 ± 0.30 μm for linear and -3.17 ± 0.37 μm for perpendicularly interlaced trajectories on the AISI 304 L surface and maximum -1.98 for linear and -2.33 for perpendicularly interlaced trajectories on the Ti6Al7Nb surface), meaning that the valleys were deeper than the peaks (also confirmed by the comparison of S_p (606.57 ± 24 μm) and S_v (618.63 ± 32 μm)) [11]. A negative skewness value, indicating the dominance of valleys over peaks, may play a critical role in functional applicability, as it implies a reduced presence of fragile surface peaks that could otherwise form stress-concentrating weak points. A surface characterized by deeper valleys and more voluminous features, therefore, suggests an improved ability to withstand external loading and enhanced wear resistance [10, 20]. This behavior can be advantageous, for example, when roughening functional components intended for the application of wear-resistant surface coatings [21]. Surfaces generated by the PWJ method under the mentioned traverse speeds could provide favorable adhesive conditions due to the formation of macrostructures (hundreds of micrometers in scale) composed of deeper valleys relative to the peaks (negative S_{sk} mentioned above) and dense surface texture distribution (maximum $S_{ku} = 9.24 \pm 0.48$ μm for linear and 9.53 ± 0.38 μm for perpendicularly interlaced trajectories > 3) [10].

With increasing traverse speeds of $v = 3$, 4, and $5\ \text{mm}\cdot\text{s}^{-1}$ in the linear trajectory, a significant reduction in measured S_a values was observed. In this case, Ti6Al4V exhibited higher surface roughness than AISI 304 L. This opposite trend compared to $v = 1$ and $2\ \text{mm}\cdot\text{s}^{-1}$ can be explained by the aforementioned material properties. In harder materials such as Ti6Al4V, the incubation period (i.e., the minimum loading time or stress required to initiate material loss) is shorter due to localized cyclic loading compared to the more ductile AISI 304 L. Another explanation could be that the incubation period for Ti6Al4V was shortened due to the easier removal of the β-phase, which required less energy to damage the material. This phenomenon was also confirmed in [18], where the authors described facilitated subsurface material removal, particularly in the regions surrounding the softer β-phase.

This indicates that, even at higher traverse speeds, the material's temporal loading remains sufficient to induce surface erosion. Conversely, AISI 304 L demonstrates better resistance to erosion at higher traverse speeds under the same temporal loading as Ti6Al4V. The onset of the material removal phase is delayed due to its superior plasticity, which provides greater resistance to cyclic loading. Therefore, it can be stated that Ti6Al4V undergoes erosion more continuously and smoothly compared to AISI 304 L, for which it is necessary to exceed a critical stress threshold before a rapid and uncontrolled rate of material removal occurs. The initiation of a more uniform erosion progression suggests that Ti6Al4V, due to its mechanical properties, also exhibits the potential for strain hardening [18]. An increase in the degree of hardening by 25.3% compared to the untreated sample was observed after linear trajectory processing at a traverse speed of $v = 5$ mm·s^{-1}. Subsurface hardening was also confirmed in [20], where it was further determined that such a surface possesses improved resistance to mechanical damage.

In the case of the perpendicularly interlaced trajectory, a high degree of erosion was also observed at traverse speeds of $v = 2$ and 4 mm·s^{-1}. Despite identical exposure times, different S_a values were recorded. For Ti6Al4V, a significant reduction in S_a was observed at all traverse speeds of $v = 6$, 8, and 10 mm·s^{-1}. This result agrees with [14], where the perpendicular interlaced trajectory produced lower surface roughness R_a compared to the linear trajectory. Conversely, for AISI 304 L, an increase in S_a was observed at traverse speeds of $v = 6$ and 8 mm·s^{-1}. It is important to note that the perpendicularly interlaced trajectory produces a surface with a very low degree of directional texture, i.e., an isotropic surface. According to the study [13], the S_{tr} parameter was consistently higher for the perpendicularly interlaced trajectory ($S_{tr} = 0.82 \pm 0.04$) compared to the linear trajectory ($S_{tr} = 0.78 \pm 0.04$), and in all cases exceeded 0.5, which indicates strong isotropy. This result implies that the generated topography exhibits uniform characteristics in all directions, suggesting more consistent behavior in terms of wear, stress distribution, or adhesion performance [22]. These results clearly demonstrate the significance of selecting between linear and perpendicular interlaced modes during PWJ surface modification. Additionally, this process helps truncate sharp peaks, potentially improving wear resistance.

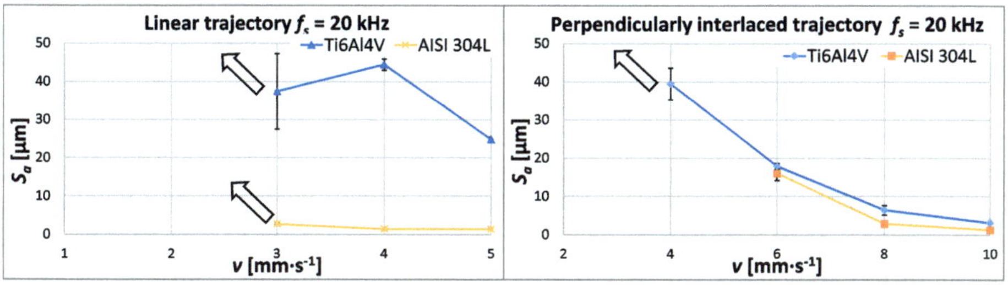

Fig. 3. Comparison of S_a surface roughness parameters at $f_S = 20$ kHz for linear and perpendicular interlaced trajectory on Ti6Al4V and AISI 304 L.

3.2 Modified Surfaces of Ti6Al4V and AISI 304 L at $f_s = 40$ kHz

Table 4 presents the average measured values of the surface roughness parameter S_a together with the calculated standard deviation for the surfaces of Ti6Al4V and AISI 304 L materials at a frequency $f_s = 40$ kHz. Qualitative analysis of the results is discussed in the next paragraph.

Table 4. Average measured roughness of S_a on Ti6Al4V and AISI 304 L at $f_s = 40$ kHz

	Linear trajectory					Perpendicularly interlaced trajectory				
v [mm·s^{-1}]	1	2	3	4	5	2	4	6	8	10
	Ti6Al4V									
S_a [µm]	9.17 ±2.46	5.23 ±0.80	1.75 ±0.22	1.36 ±0.01	2.96 ±0.25	13.74 ±3.21	5.02 ±0.43	2.33 ±0.45	2.69 ±0.17	1.73 ±0.09
	AISI 304 L									
S_a [µm]	25.56 ±1.87	8.77 ±0.81	5.65 ±0.23	2.04 ±0.06	1.37 ±0.02	16.25 ±1.89	2.65 ±0.21	1.40 ±0.01	1.23 ±0.01	1.04 ±0.02

The frequency $f_s = 40$ kHz caused significantly different erosion behavior of the PWJ compared to $f_s = 20$ kHz, while maintaining identical process parameters and traverse speeds. The phenomenon of reduced erosion at higher excitation frequency was also observed in [23]. At $f_s = 40$ kHz, the material is subjected to twice the periodic loading, thereby increasing fatigue by shortening the time available for mechanical relaxation. Under the same hydraulic conditions ($p = 70$ MPa, $d = 0.8$ mm), the water jet is divided into twice as many smaller water clusters impacting the target surface. Consequently, this reduces the impact pressure intensity, thereby altering the material's erosion response. For the Ti6Al4V alloy, differences between the 20 and 40 kHz frequencies were observed: 86–96% for the linear trajectory and 44–87% for the perpendicular interlaced trajectory, with lower roughness values at 40 kHz in all cases. For AISI 304 L steel, the differences ranged from 0% to 108.5% for the linear trajectory and 16–95% for the perpendicular interlaced trajectory. The results clearly indicate that significantly greater variations in roughness were observed between 20 and 40 kHz for AISI 304 L than for Ti6Al4V. In certain cases, such as sandstone erosion [21], higher erosion efficiency was observed at $f_s = 40$ kHz than at $f_s = 20$ kHz. Therefore, it is evident that the change in erosion potential with excitation frequency is strongly dependent on the type and structure of the target material.

This was also confirmed by results comparing the surface modifications of Ti6Al4V and AISI 304 L, where, at $f_s = 40$ kHz, higher erosion was observed for AISI 304 L than for Ti6Al4V (Fig. 4). The surface modification results at $f_s = 40$ kHz suggest that the average measured roughness parameters could be suitable for surface modification of biomaterials used in medical applications. This assumption is based on comparisons with implant surface roughness, where typical S_a values range from 1 to 10 μm [1] depending on implant type and function. Achieving optimal surface roughness may also shorten the time required for primary fixation and improve biological stability [24]. The PWJ modification process enables the creation of rough surfaces, with erosion level controlled by adjusting the processing strategy or traverse speed. In this way, it is possible to achieve desired surface roughness characteristics (S_a) tailored to specific applications. According to [9, 25], implant surfaces are categorized into several groups (smooth, minimally rough, moderately rough, and rough), each offering different application suitability (e.g., dental or hip implants) depending on bone structure [26]. Based on this classification, it can be concluded that the PWJ method can generate moderately rough and rough surfaces on Ti6Al4V and AISI 304 L, depending on the traverse speed. Specifically, for linear trajectories, moderately rough surfaces with roughness values between $S_a = 1$–2 μm were achieved at traverse speeds of $v = 3$, 4, and 5 mm·s^{-1} on both Ti6Al4V and AISI 304 L. For perpendicular interlaced trajectories, moderately rough surfaces were also generated at speeds of $v = 6$, 8, and 10 mm·s^{-1}, depending on the material type. Such surfaces could therefore be suitable for modification of hip implant stems for cementless fixation, where surface roughness around 3 μm [27], or up to 10 μm [28, 29], is typically used. Rough surfaces were achieved at traverse speeds of $v = 1$ and 2 mm·s^{-1} for the linear trajectory and $v = 4$ mm·s^{-1} for the perpendicular trajectory on Ti6Al4V, and at $v = 2$ and 3 mm·s^{-1} for the linear trajectory and $v = 4$ mm·s^{-1} for the perpendicular trajectory on AISI 304 L. At these roughness levels, enhanced active cell cultivation on the surface can be expected, increasing over time after implantation [30]. Moreover, even at 40 kHz, a high degree of isotropy was observed, approximately 72.83% for the linear trajectory and 96.74% for the orthogonal oscillating strategy [10]. This further confirms the topography's lack of directional dependence, which may facilitate uniform cultivation of bone cells on the surface. However, to clearly determine the suitability of PWJ-modified surfaces, additional in vitro studies would be required to verify their potential for biomedical use.

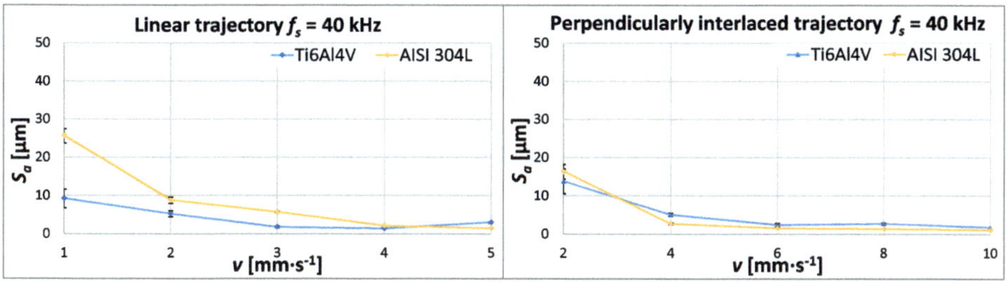

Fig. 4. Comparison of S_a surface roughness parameters at $f_S = 40$ kHz for linear and perpendicular interlaced trajectory on Ti6Al4V and AISI 304 L.

3.3 Limitations and Future Research Directions

The results of this study provide a fundamental overview of the possibilities for modifying Ti alloys and AISI steels. Distinct erosion characteristics were observed depending on the materials' mechanical properties, which determined their surface roughness parameter S_a and thus their potential fields of application. Based on the results of this study, a clear table comparing dominant phenomena at frequencies of $f_s = 20$ and 40 kHz was created (Table 5).

Table 5. Comparison of dominant phenomena in surface modification using frequencies $f_S = 20$ and 40 kHz

Parameter	$f_S = 20$ kHz	$f_S = 40$ kHz
Observed behavior	AISI 304 L erodes more at low speed; Ti6Al4V erodes more at high speed.	Clear inversion: AISI 304 L erodes significantly more at all speeds.
Relative energy	High	Low
Erosion regime	Erosion is governed by resistance to gross plastic deformation.	Erosion is governed by resistance to damage accumulation from cyclic plastic deformation.
Key characterization	High hardness and strength (favors Ti6Al4V)	High fatigue strength and resistance to cyclic strain localization (favors Ti6Al4V)
Key property promoting erosion	Low hardness and strength (promotes erosion in AISI 304 L)	High ductility and capacity for plastic deformation (promotes fatigue damage in AISI 304 L)

However, several factors remain unclear and could further clarify both the technology's practical applicability and the behavior of the modified surface. From a technological standpoint, the experiments were mainly focused on surface modification under atmospheric conditions. In this case, fluctuations in water pressure within the acoustic chamber manifest as variations in jet velocity after exiting the nozzle, and the coherent water stream is subsequently divided into discrete segments in the air. Nevertheless, study [11] has shown that this method can also modify surfaces under submerged conditions. This introduces a new perspective on using PWJ as a tool for underwater surface treatment aimed at roughening or cleaning. Submerged surface cleaning could be applied to maintain pillar structures or ship hulls to prevent the spread of marine organisms that can contaminate harbors [31]. Currently, such processes are performed using mechanical removal or continuous water jet (CWJ) methods [32]. The application of the PWJ technique could reduce the operational pressure required during cleaning, thereby lowering energy costs. However, further research is needed to determine how the surrounding static water affects the formation of water segments and, consequently, their erosion potential. In submerged conditions, rapid changes in the velocity of the entrained static water can also induce cavitation bubbles [33]. These bubbles may participate in the erosion process, influencing both the erosion potential and the resulting surface structure. Therefore, a detailed analysis of submerged surface modification processes could clarify whether the erosion potential is reduced by the surrounding static water or enhanced by potential cavitation effects. It has also been shown that grain size variations resulting from heat treatment significantly influence the erosion behavior of the target surface [34]. Therefore, future research could focus on modifying these materials after heat treatment (annealing) to characterize the resulting topographical differences compared to the base state and potentially propose new approaches to surface formation.

Another research gap concerns the geometry of the modified surface itself. In all comparative studies, the target materials had flat surfaces, with water segment impacts oriented perpendicularly (90°) to the surface, utilizing the full erosion potential. However, for potential implant applications, components often exhibit circular or partially curved geometries. The impact of liquid on a curved surface may reduce the effective erosion potential, as the central part of the water segment impacts the material perpendicularly, while the lateral parts experience shear along the curved surface. The portion of the water segment that effectively contributes to erosion versus that which transitions into shear flow depends on the ratio of the segment diameter "D" to the surface curvature radius "r". Consequently, the next essential step would be to investigate PWJ's ability to modify curved surfaces with varying radii of curvature.

4 Conclusion

This study compares the surface modification behavior of Ti6Al4V alloy and AISI 304 L stainless steel using the PWJ method under identical experimental conditions. The original contribution of the work lies in revealing a frequency-dependent inversion in the erosion behavior of two materially distinct alloys. The findings demonstrated that by changing the impact frequency of water pulses, it is possible to radically alter which material exhibits greater material removal. This insight provides a deeper understanding of the interaction between PWJ parameters and the mechanical properties of materials, opening new possibilities for targeted and controlled creation of surface structures. The key findings supporting this contribution are summarized as follows:

- At a frequency of $f_s = 20$ kHz, the Ti6Al4V alloy exhibited a higher erosion rate, whereas increasing the frequency to $f_s = 40$ kHz reversed this trend, with significantly higher erosion observed on AISI 304 L steel.

- The differing material responses were attributed to their mechanical properties. It was shown that a material's ability to absorb impact energy through plastic deformation is a decisive factor. Ti6Al4V undergoes continuous erosion, while AISI 304 L requires exceeding a critical load threshold before rapid material removal occurs.
- By careful selection of frequency, traverse speed, and trajectory, surfaces with defined roughness can be generated on both materials. Depending on the parameters, surfaces ranging from moderately rough to rough, suitable for implantology applications can be achieved.
- The comparison between linear and perpendicular interlaced trajectories showed that the perpendicular mode results in a more homogeneous and isotropic surface structure, which is essential for applications requiring uniform adhesive properties.

Acknowledgement. This study was financially supported by the Slovak Research and Development Agency under Contract No. APVV-22-0391. The authors are also grateful to Dr. Josef Foldyna, for permission to conduct the experiments using PWJ Technology. The experiments were conducted at the Institute of Geonics of the Czech Academy of Sciences, Ostrava-Poruba, Czech Republic, with the support for the long-term conceptual development of the research institution RVO: 68145535.

References

1. Krishna Alla, R., Ginjupalli, K., Upadhya, N., Shammas, M., Krishna Ravi, R., Sekhar, R.: Surface roughness of implants: a review. Trends Biomat. Artif. Organs. **25**(3), 112–118 (2011)
2. Vilar, R.: Laser surface modification of biomaterials: techniques and applications (2016)
3. Arola, D.D., McCain, M.L.: Abrasive waterjet peening: a new method of surface preparation for metal orthopedic implants. J. Biomed. Mater. Res. **53**, 536–546 (2000)
4. Srivastava, M., et al.: Ultrasonically generated pulsed water jet peening of austenitic stainless-steel surfaces. J. Manuf. Process. **32**, 455–468 (2018). https://doi.org/10.1016/j.jmapro.2018.03.016
5. Hloch, S., et al.: Effect of pressure of pulsating water jet moving along stair trajectory on erosion depth, surface morphology and microhardness. Wear, 452–453 (2020). https://doi.org/10.1016/j.wear.2020.203278
6. Nag, A., Stolarik, G., Hloch, S.: Comparison of the effects of standard nozzles and extension tubes on the erosion of Pmma using pulsating water jet technology. MM Sci. J. **2024**, 7687–7694 (2024). https://doi.org/10.17973/MMSJ.2024_11_2024099
7. Stolárik, G., Nag, A., Petrů, J., Svobodová, J., Hloch, S.: Ultrasonic pulsating water jet peening: influence of pressure and pattern strategy. Materials. **14** (2021). https://doi.org/10.3390/ma14206019
8. Srivastava, M., Hloch, S., Krejci, L., Chattopadhyaya, S., Dixit, A.R., Foldyna, J.: Residual stress and surface properties of stainless steel welded joints induced by ultrasonic pulsed water jet peening. Measurement (Lond). **127**, 453–462 (2018). https://doi.org/10.1016/j.measurement.2018.06.012
9. Parithimarkalaignan, S., Padmanabhan, T.V.: Osseointegration: an update (2013). https://doi.org/10.1007/s13191-013-0252-z.
10. Stolárik, G., Klichová, D., Poloprudský, J., Nag, A., Hloch, S.: Assessment of surface irregularities created by controlled liquid droplet on the surface of stainless steel AISI 304L. Engineering science and technology, an. Int. J. **47** (2023). https://doi.org/10.1016/j.jestch.2023.101558

11. Stolárik, G., Klichová, D., Poloprudský, J., Chlupová, A., Nag, A., Hloch, S.: Submerged surface texturing of AISI 304L using the pulsating water jet method. Arch. Civil Mech. Eng. **24**, 207 (2024). https://doi.org/10.1007/s43452-024-01029-x
12. Stolárik, G., Svobodová, J., Klichová, D., Nag, A., Hloch, S.: Titanium surface roughening with ultrasonic pulsating water jet. J. Manuf. Process. **90** (2023). https://doi.org/10.1016/j.jmapro.2023.02.013
13. Klichová, D., et al.: Surface modification of titanium alloy Ti6Al7Nb by high-frequency droplet impingement. Wear. **570** (2025). https://doi.org/10.1016/j.wear.2025.205896
14. Klichova, D., et al.: Utilising of water hammer effect for surface roughening of Ti6Al4V. Int. J. Adv. Manuf. Technol. **126**, 5633–5647 (2023). https://doi.org/10.1007/s00170-023-11521-y
15. Nag, A., Stolárik, G., Svehla, B., Hloch, S.: Effect of water flow rate on operating frequency and power during acoustic chamber tuning. In: International Conference on Manufacturing Engineering and Materials, pp. 142–154. Springer International Publishing, Cham (2021)
16. Srivastava, M., Nag, A., Chattopadhyaya, S., Hloch, S.: Standoff distance in ultrasonic pulsating water jet. Materials. **14**, 1–18 (2021). https://doi.org/10.3390/ma14010088
17. Petru, J., Skorepa, R., Nag, A., Stolarik, G., Hloch, S.: Evaluation of erosion resistance of additive 3d printed materials under continuous and pulsating water jet exposure. MM Sci. J., 8162–8168 (2025). https://doi.org/10.17973/MMSJ.2025_03_2025006
18. Siahpour, P., et al.: Surface characteristics and residual stress generation in Ti-6Al-4 V following ultrasonic pulsed water jet peening. Surf. Coat. Technol. **445** (2022). https://doi.org/10.1016/j.surfcoat.2022.128691
19. Siahpour, P., Amegadzie, M.Y., Tieu, A., Donaldson, I.W., Plucknett, K.P.: Ultrasonic pulsed waterjet peening of commercially-pure titanium. Surf. Coat. Technol. **472** (2023). https://doi.org/10.1016/j.surfcoat.2023.129953
20. Siahpour, P., et al.: Evaluating surface mechanical properties and wear resistance of Ti–6Al–4V alloy subjected to ultrasonic pulsed waterjet peening. Wear, 550–551 (2024). https://doi.org/10.1016/j.wear.2024.205400
21. Croll, S.G.: Surface roughness profile and its effect on coating adhesion and corrosion protection: a review. Prog. Org. Coat. **148**, 105847 (2020)
22. Kubiak, K.J., Wilson, M.C.T., Mathia, T.G., Carval, P.: Wettability versus roughness of engineering surfaces. Wear. **271**, 523–528 (2011). https://doi.org/10.1016/j.wear.2010.03.029
23. Nag, A., Hvizdos, P., Dixit, A.R., Petrů, J., Hloch, S.: Influence of the frequency and flow rate of a pulsating water jet on the wear damage of tantalum. Wear. **477** (2021). https://doi.org/10.1016/j.wear.2021.203893
24. Romero-Serrano, M., Romero-Ruiz, M.-M., Herrero-Climent, M., Rios-Carrasco, B., Gil-Mur, J.: Correlation between implant surface roughness and implant stability: a systematic review. Dent J. (Basel). **12**, 276 (2024). https://doi.org/10.3390/dj12090276
25. Periospot: ARE ALL THE IMPLANT SURFACES THE SAME?
26. Bergkvist, G.: Kwang-Joon Koh: bone density at implant sites and its relationship to assessment of bone quality and treatment outcome. Int. J. Oral Maxillofacial Implants. **25**, 321–328 (2010)
27. Gielen, A.M.C., et al.: Utilizing biomaterial surface properties to improve orthopedic hip implant safety and function in a safe-by-design approach. Front. Bioeng. Biotechnol. (2025). https://doi.org/10.3389/fbioe.2025.1504883
28. Verdonschot, N., Tanck, E., Huiskes, R.: Effects of prosthesis surface roughness on the failure process of cemented hip implants after stem-cement debonding. J. Biomed. Mater. Res. **42**, 554–559 (1998)
29. Suzuki, K., Aoki, K., Ohya, K.: Effects of surface roughness of titanium implants on bone remodeling activity of femur in rabbits. Bone. **21**(6), 507–514 (1997)

30. Andrukhov, O., et al.: Proliferation, behavior, and differentiation of osteoblasts on surfaces of different microroughness. Dent. Mater. **32**, 1374–1384 (2016). https://doi.org/10.1016/j.dental.2016.08.217
31. Song, C., Cui, W.: Review of underwater ship Hull cleaning technologies: C. Song, W. Cui. J. Mar. Sci. Appl. **19**(3), 415–429 (2020)
32. Albitar, H., Dandan, K., Ananiev, A., Kalaykov, I.: Underwater robotics: surface cleaning technics, adhesion and locomotion systems. Int. J. Adv. Robot. Syst. **13** (2016). https://doi.org/10.5772/62060
33. Soyama, H.: Cavitation peening: a review. Meta. **10**(2), 270 (2020). https://doi.org/10.3390/met10020270
34. Chlupova, A., Poloprudsky, J., Nag, A., Stolarik, G.: Hydrodynamic erosion of copper with different grain sizes. MM Sci. J. **2025** (2025). https://doi.org/10.17973/MMSJ.2025_10_2025097

Comparative Mechanical and Tribological Analysis of 3D Printed Harmonic and Cycloidal Gearboxes

Paul Ciudin[1], Marius Pustan[1], Florin Popişter[1](✉), Friedemann Schaber[2], Mihai Steopan[1], and Horea Ştefan Goia[1]

[1] Technical University of Cluj Napoca, Cluj Napoca, Romania
florin.popister@muri.utcluj.ro
[2] University of Northampton, Northampton, UK

Abstract. Harmonic and cycloidal gear transmissions are widely used in applications where compact size and high reduction ratios matter, except the two mechanisms rely on quite diverse ways of sharing loads and creating motion. To explore how these differences play out in practice, especially when friction dominates their behaviour, two polyethylene printed prototypes were designed and assessed under identical operating conditions. The study examines both designs during several running cycles at increasing input speeds, first with reduced internal lubrication and later with a light increase of lubrication applied. Temperature and sound were tracked throughout each experiment using a thermal camera and a sound level meter placed at a fixed position relative to the test bench. The results show clear contrasts between the two reducers. The harmonic unit tends to heat up rapidly, while the cycloidal model produces a distinct pulsating noise pattern. Lubrication moderates both effects, but not to the same extent. These observations highlight how mechanical layout and material properties shape the tribological performance of additively manufactured reducers.

Keywords: Harmonic Drive · Cycloidal Drive · Tribology · Comparative Analysis · Acoustic Monitoring · Thermal Behaviour

1 Introduction

Gear transmissions capable of producing high reduction ratios in a compact envelope have become essential in fields such as robotics, light automation, and precision equipment. Among the various solutions available, harmonic and cycloidal drives occupy a particular place in which both can achieve large reductions in a single stage, also they do so through mechanisms that behave quite differently under load as compared in literature [1]. Harmonic drives depend on controlled elastic deformation of a thin metallic cup, which allows a high tooth count difference in a small housing, while cycloidal drives depend on an eccentric motion and multiple rolling contacts to distribute forces [2]. Their widespread use has less to do with tradition and more with the practical benefits these designs offer high torque density, reduced backlash, repeatable motion, and the ability to integrate easily into compact assemblies [3].

M. Hryb et al. (Eds.): MANUFACTURING 2026, LNME, pp. 172–186, 2026.
https://doi.org/10.1007/978-3-032-23038-6_13

Despite these strengths, both mechanisms face the same fundamental challenges present in any gearbox, like friction, heat generation, and wear. Tribological behaviour can strongly influence efficiency, lifespan, and noise emissions [1]. In miniature transmissions these effects are even more pronounced because small clearances magnify sensitivity to surface roughness, lubrication quality, and material stiffness [4]. For harmonic drives, sliding contact along the tooth flanks and the continuous flexing of the compliant element can create localised heating and fatigue as related damage [5]. Cycloidal reducers, while more tolerant to shock loads, tend to experience intermittent impacts and mixed rolling sliding conditions at the pin interfaces, which may produce distinct acoustic signatures and abrasion [6]. Understanding these interactions is usually straightforward with industrial grade steel components but far less discussed when the same mechanisms are reproduced using additive manufacturing with polyethylene or Fiber reinforced polymers [7].

From a manufacturing sustainability perspective, fused deposition modelling (FDM) has enabled rapid and resource-efficient prototyping of complex gear transmissions, allowing design concepts to be evaluated without the material waste and energy demand associated with conventional machining. Nevertheless, printed components introduce their own specific constraints. Polymers such as glycol-modified copolyesters exhibit higher interfacial friction than metals, and the layer-by-layer deposition process produces surface textures that differ notably from those of machined teeth [8]. Dimensional deviations can further alter intended contact geometries. Alternately than being treated as limitations, these characteristics amplify tribological effects, making frictional losses, heat generation, and noise development easier to detect during testing.

Studies which address harmonic and cycloidal drives individually, direct experimental comparisons using low-cost, non-invasive diagnostic methods remain limited. The present work addresses this gap by investigating two 3D-printed reducers operating under identical conditions. The objective is to compare their mechanical configuration, tribological response, temperature evolution, and acoustic signatures during controlled running cycles. By combining additive manufacturing with thermal and acoustic monitoring, the study supports a quality-oriented and sustainable design approach, enabling early identification of unfavourable mechanical behaviours before transitioning to energy and machining production methods.

2 Mechanical Design Characteristics

2.1 Harmonic Drive Characteristics

A harmonic drive operates on a different principle than most conventional gear systems. Instead of relying on a rigid set of teeth meshing in a predictable pattern, it makes use of a flexible component, which is called a flex spline, which is forced into an elliptical shape by a rotating wave generator. As the wave generator turns, the flex spline bends outward along two opposing lobes, allowing its teeth to engage with those of a rigid circular spline in only two regions at a time [9]. Because the flex spline has slightly fewer teeth, each rotation of the wave generator advances the output by a small but precise amount, producing the high reduction ratio. In this study, the printed model provides a ratio of

1:50. Even if the mechanism appears complex, the actual motion is continuous, and the flexing behaviour is fundamental to its performance [10].

Harmonic drives are often chosen when precision is more important than brute strength. The main mechanical strengths are the exceptionally low backlash, high positional repeatability, and the ability to achieve large ratios in a single stage, which have made them a popular choice in robotic joints and optical systems. The almost complete elimination of play between teeth gives these drives a characteristic smoothness that is difficult to match with classical spur or planetary gears [3].

Nevertheless, the same qualities that give harmonic drives their accuracy also introduce certain mechanical vulnerabilities. The flex spline undergoes continuous elastic deformation, as seen from the drive construction in Fig. 1, which concentrates stress along specific regions of the cup. When external loads vary or exceed design expectations, these stresses can grow sharply, making the mechanism sensitive to overload and long-term fatigue. Tooth engagement is also not uniform around the circumference; only the two elliptical lobes carry the transmitted load, which creates localised regions of high contact pressure [11].

Fig. 1. 3D Printed Harmonic Drive.

From a tribological standpoint, the harmonic drive tends to operate in conditions dominated by sliding rather than rolling. As the flex spline teeth sweep into and out of engagement, the surfaces experience small but repeated sliding motions that encourage adhesive wear as discussed in literature [12]. Over time, especially in plastic prototypes such as polyethylene prints, this can lead to local heating, material smearing, or surface fatigue [13]. These tendencies make lubrication important, even though the design can technically run dry in demonstration setups. In practice, the harmonic drive behaviour is closely tied to how well the contacting surfaces are managed, both in terms of material properties and lubrication quality.

2.2 Cycloidal Drive Characteristics

A cycloidal reducer transmits motion through a mechanism which is driven by an eccentrically mounted input shaft that induces a controlled orbital motion of the cycloidal disc. This disc follows a constrained, non-circular trajectory relative to the fixed ring of pins, resulting in a wobbling yet precisely guided movement that enables speed reduction [6]. As the central disc moves, its lobes trace a trochoidal curve and interact with a ring of fixed pins arranged around the housing [14]. In reaction, a second set of pins mounted on the output plate follows the motion of the cycloidal profile, converting the eccentric motion into a slow, steady rotation [15].

Several lobes of the reducer engage with several pins at once, forces are distributed over larger areas compared to traditional spur gears. This makes the design inherently resistant to sudden shocks and suited for tasks where motion is not perfectly fluid [16]. The reduction ratios achievable in a single stage are also considerably often comparable to those of harmonic drives while still maintaining acceptable backlash levels. In a printed prototype, some of these advantages remain visible, even if not to the same degree as in machined steel versions.

The eccentric input illustrated in Fig. 2, creates a characteristic pulsating motion that shows up as slight speed fluctuations at the output, especially when the components are not manufactured with tight tolerances. These pulsations are usually harmless but tend to introduce vibration and noise. The mechanism also depends on precise geometry between the cycloidal lobes and the ring pins. Dimensional errors, such as those introduced by FDM printing, can amplify impacts or misalignment [17].

Fig. 2. 3D Printed Cycloidal Drive.

From a tribological perspective, the cycloidal drive typically experiences alternating zones of rolling contact, partial sliding, and brief shock like interactions when the lobes transition between pins. In metal components this behaviour is controlled, but in polyethylene terephthalate glycol parts the rough surface and limited stiffness make abrasion and micro-chipping more likely [18]. The intermittent nature of the contact path also tends to create a distinct acoustic pattern more rhythmic and pulsating than the continuous tone of a harmonic reducer. Lubrication can soften these effects, but the interplay between rolling and sliding remains characteristic of the cycloidal mechanism [19].

3 Experimental Setup and Procedure

3.1 3D-Printed Gearboxes

The reducers were manufactured from polyethylene terephthalate glycol modified polymer processed through fused deposition modelling, a material known for moderate stiffness and high ductility. Typical mechanical properties reported for PETG include a tensile strength of approximately 40 MPa and a Young's modulus close to 1.8 GPa [20], values that are lower than those of metallic transmission components and therefore promote increased elastic compliance under contact loading.

For a layer height of 0.2 mm, PETG parts produced by fused filament deposition typically result in rough surfaces, originating from the layer deposition process and filament geometry, as also discussed in literature [21]. Such surface conditions favour boundary dominated lubrication regimes, increase real contact area, and intensify frictional interactions. In the present study, identical printing parameters and orientation were applied to both reducers, ensuring that material and surface effects function as a common baseline. The elevated thermal response and pronounced acoustic emissions observed experimentally are consistent with these tribological conditions and highlight how additively manufactured polymer surfaces amplify friction and impact related phenomena during operation. Both reducers were produced using the same printing parameters illustrated in Table 1 below and a similar filament batch of distinct colours.

Table 1. 3D printing parameters.

Parameter	Value
Printer model	Ender 3v3 KE
Layer height	0,20 mm
Nozzle diameter	0,40 mm
Extrusion temperature	240 °C
Bed temperature	80 °C
Infill density	40% (grid)
Print speed	50 mm/s
Cooling fan	60% above first layers
Shell thickness	1,2 mm

The testing procedure was conducted in two phases. In the initial stage, both reducers were assembled and operated with low internal lubrication. Running the assemblies with reduced lubrication helped reveal how the printed surfaces interact under load and accentuated the characteristic differences in frictional response between the two mechanical layouts. After completing these baseline trials, a thin layer of grease was introduced inside each gearbox, and the full set of tests was repeated. This second phase allowed an immediate comparison of how lubrication modifies temperature rise, noise output, and the overall motion.

3.2 Test Bench

The experimental setup presented in Fig. 3 was built around a straightforward but dependable test bench designed to keep both gearboxes under comparable conditions. Two NEMA 23 stepper motors served as the driving units with the output torque of 1.24 Nm. Each motor was connected to its respective gearbox through a metal shaft fitted with a small elastic coupling. This coupling was used to absorb minor alignment errors and to prevent sudden load changes from transferring directly into the printed components.

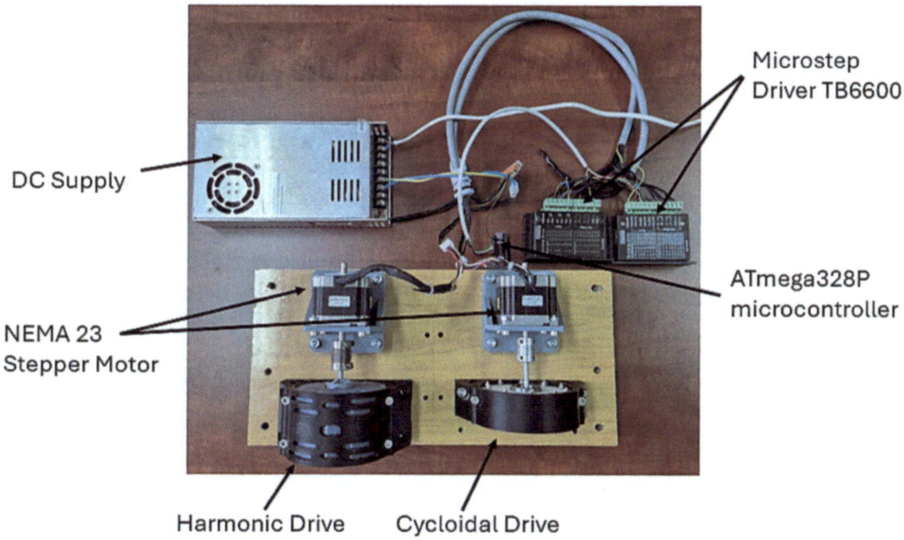

Fig. 3. Elements in the test bench composition.

Both motors and reducers were secured onto the table using custom 3D-printed mounting frames. These supports were printed with a 90% infill density to ensure rigidity and to minimize any unwanted flexing that could influence the measurements. The use of printed fixtures also made it easier to align the assemblies and allowed small adjustments without needing machine parts.

The thermal response of the gearboxes was monitored using an infrared camera positioned at an average distance of 10 cm from the external housing surfaces. All measurements were conducted under controlled laboratory conditions, with an ambient temperature of 22.5 °C, so that the recorded thermal variations could be associated primarily with internal dissipation rather than environmental fluctuations.

Acoustic data were collected with a sound level meter placed at a fixed distance of 5 cm from each reducer during operation. Prior to testing, the background noise in the laboratory was measured and remained close to 52 dB for the duration of the experiments. Maintaining a constant sensor position allowed the recorded sound pressure levels to reflect the dynamic response of the transmission mechanisms themselves, ensuring comparability between the two reducer configurations.

3.3 Experimental Procedure and Operating Sequence

At the start of each test, the two reducers were allowed to reach ambient thermal equilibrium to avoid any residual heating influencing the initial readings. For each reducer, measurements were conducted at all input speeds investigated in the study, ranging from 100 to 600 rpm. At each speed level, temperature and sound pressure data were recorded continuously over defined operating intervals. The values reported in the study represent mean values derived from these measurements.

Temperature measurements were acquired using a PWBKA 4 A1 thermal imaging camera, positioned to provide a stable view of the reducer housings. The camera was kept at a fixed distance and angle, ensuring that the coverage of the thermal frames remained consistent across all speed levels. Acoustic measurements were taken with a PCE sound level meter, a handheld device commonly used for machinery diagnostics and environmental noise assessments.

To minimise friction and stabilise the contact conditions during the lubricated tests, both reducers were greased with Super Lube Multi-Purpose Synthetic Grease with Syncolon illustrated in Fig. 4, a PTFE fortified lubricant well suited for polymer-based mechanisms. According to its technical specification, the grease operates reliably between 43 °C and 232 °C, has a viscosity of 69 cSt at 40 °C, and has an NLGI Grade 2 consistency, even when the printed components warm up [22].

Fig. 4. Super Lube Multi-Purpose Synthetic Grease [22].

The experimental setup was conducted under unloaded operating conditions, as the primary objective of the study was to isolate and compare the mechanical and tribological behaviour of the two reducer architectures. Operating without external load eliminates variability associated with output torque, alignment, and load-dependent deformation, allowing internal losses related to surface interaction, elastic compliance, and kinematic constraints to reflect in the response. In additively manufactured reducers produced by fused deposition modelling, internal resistance is strongly influenced by the characteristic surface roughness of printed parts, which promotes mixed sliding–rolling contact even in nominally rolling interfaces. Backlash related effects, amplified by printing tolerances and surface asperities, which are reflected in the measured noise patterns. This indirect but energy consistent approach provides a robust comparative basis for assessing internal dissipation mechanisms under identical unloaded conditions.

4 Comparative Results and Analysis

4.1 Thermal Performance

The thermal measurements obtained for the two reducers reveal a clear divergence in how each mechanism responds as the input speed increases. The thermographic frames in Fig. 5 show that the harmonic reducer experiences a steady and quite noticeable rise in surface temperature over the full speed range. Starting from approximately 25 °C at 100 rpm, the temperature climbs progressively and reaches values around 36 °C once the input reaches 600 rpm. The images make it evident that the warming is concentrated around the flex spline region, where the cyclic deformation and sliding contact generate localised frictional losses. As speed increases, these losses accumulate more rapidly, which explains the sharp gradient visible between 300 rpm and 500 rpm.

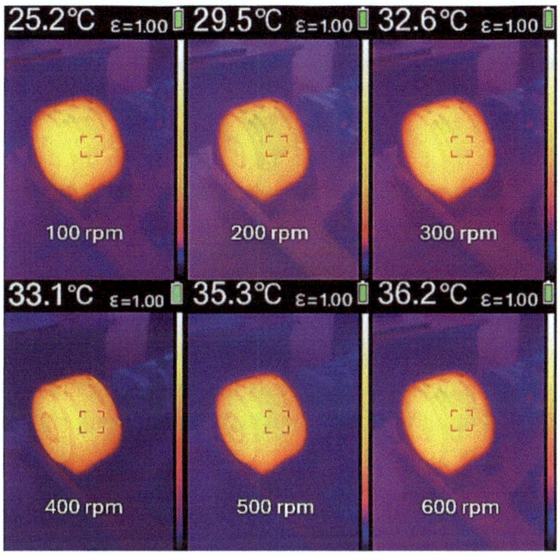

Fig. 5. Temperature test for Harmonic reducer.

By contrast, the thermal evolution of the cycloidal reducer, shown in Fig. 6, follows a milder trajectory. Although the surface temperature does increase with speed, the variation remains much smaller, from 25 °C at 100 rpm to just under 29 °C at 600 rpm. The heat footprint is also more uniform, with no strong concentration zones. This behaviour reflects the way the cycloidal mechanism distributes loads across several lobes and pins, resulting in lower peak pressures and less severe sliding at any single point.

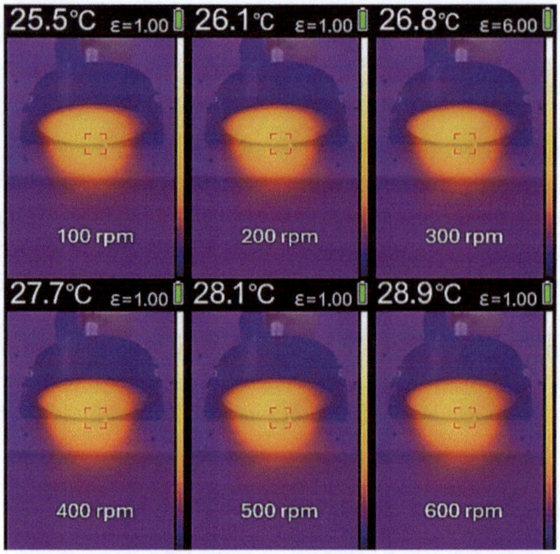

Fig. 6. Temperature test for Cycloidal reducer.

These observations become even clearer when the temperatures are plotted over time, as in Fig. 7. Both reducers were operated under identical conditions over a three-hour interval, with the input speed increased in discrete steps of 100 rpm. The harmonic unit displays a pronounced escalation during the mid-range speeds, followed by a plateau near the end of the test when the speed stabilises at 600 rpm. In contrast, the cycloidal reducer follows a far shallower curve, rising gradually but never reaching the same thermal level. The difference between the two traces suggests that the harmonic drive converts a larger portion of the mechanical input into frictional heating, particularly during the transitions between speed levels.

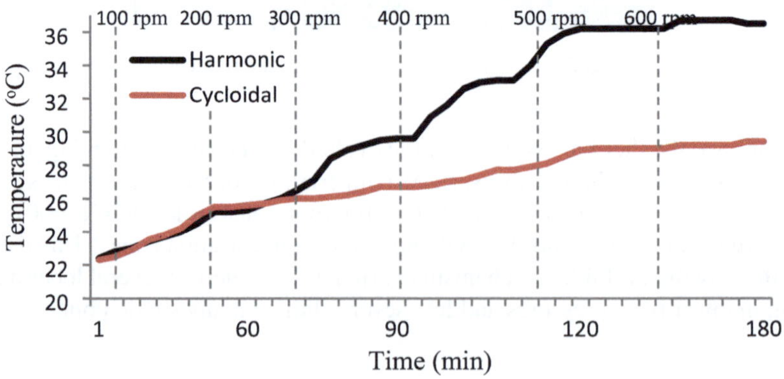

Fig. 7. Temperature evolution on the surface of the Harmonic and Cycloidal reducers over 180 min in the speed range 100–600 rpm of the input motor.

From a tribological standpoint, these results are consistent with the inherent characteristics of each drive. The harmonic reducer relies on continuous elastic deformation of the flex spline and repetitive sliding between tooth flanks, which inevitably produces more heat, especially in a material like polyethylene terephthalate glycol that has limited stiffness and a high friction coefficient. The cycloidal mechanism avoids these concentrated sliding zones, and its load-sharing geometry helps to disperse energy over multiple contacts. This explains why the temperature of the cycloidal model remained consistently lower across all speeds.

4.2 Acoustic Performance

The acoustic measurements provide another clear distinction between the two reducers and reveal how their internal kinematics manifest as airborne noise. The test configuration is illustrated in Fig. 8, where the microphone was positioned at a constant distance from each unit to maintain comparable readings. All results presented here correspond to operation with light lubrication, which was applied uniformly before testing. A second series of trials was later conducted using a noticeably thicker lubricant film, but the differences recorded were marginal and fell within the typical fluctuation range of the sound level meter as a result, the lightly lubricated condition is the most representative treatment for both reducers.

Fig. 8. Sound level tests for Harmonic and Cycloidal reducers during operation.

The noise evolution over the full three-hour interval is plotted in Fig. 8, where the sound traces mirror the speed steps imposed during the experiment. The harmonic reducer consistently produced a smoother and less erratic profile, particularly at lower and mid-range speeds. Starting around 73–75 dB at 100 rpm, the sound level increased steadily with speed and remained at 88–90 dB during the final segment at 600 rpm. The absence of sudden spikes suggests that the deformation-based tooth engagement characteristic of harmonic drives generates a continuous acoustic footprint. Even if the inner circle is repeatedly bent into the elliptical shape imposed by the wave generator, the resulting contact is distributed along gradually shifting tooth pairs, which produces a sustained but not highly impulsive noise signature.

The cycloidal reducer exhibited a markedly different acoustic pattern. From the beginning of the test, the sound readings were consistently higher and more irregular than those of the harmonic unit, starting near 77–79 dB and rising to peaks in the 92–95 dB range at higher speeds. This pulsating, wave-like pattern visible in Fig. 9 aligns with the mechanical nature of the cycloidal mechanism. These micro-impacts, occur at a frequency that depends directly on input speed, which explains the increasingly jagged appearance of the cycloidal trace as the test progresses.

The thermal images presented earlier offer indirect support for these acoustic patterns. While the harmonic reducer absorbed more frictional energy in the form of heat, its temperature rise coincided with only moderate acoustic variation. The cycloidal reducer, despite remaining cooler as in Fig. 6, converted a portion of its mechanical interactions into higher-amplitude airborne noise. This result is expected, since rolling sliding contacts combined with intermittent load transfer tend to generate stronger acoustic emissions than continuous sliding, particularly in polymer-based systems where surface irregularities intensify impact events.

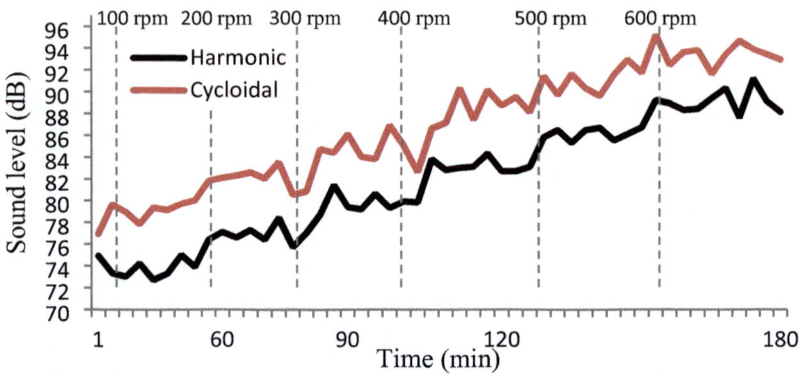

Fig. 9. Sound level measurement of the Harmonic and Cycloidal reducers over 180 min in the speed range 100–600 rpm of the input motor.

Increasing the lubrication level did not alter these tendencies in a meaningful way. Although the thicker grease layer reduced some of the sharpest peaks for the cycloidal reducer and slightly softened the harmonic unit tone, neither mechanism experienced a substantial shift in its overall acoustic profile. The cycloidal drive retained its characteristic pulsating noise, while the harmonic reducer continued to produce a steadier, more uniform sound field. This limited improvement suggests that the sound generation mechanisms in both reducers are governed primarily by geometric factors and contact kinematics rather than by the lubricant volume alone.

4.3 Summary of the Main Mechanical and Tribological Differences Based on the Test Results

The combined results show that the two reducer architectures behave differently under identical operating conditions, despite being produced from the same polymer and printed with identical parameters. The harmonic drive displayed higher average temperatures during all speed intervals, stabilizing near 36 °C at 600 rpm, whereas the cycloidal mechanism remained comparatively cooler, reaching only 28–29 °C in the same conditions. In contrast, the acoustic data showed the opposite trend that the harmonic unit produced a more uniform noise level, averaging 86–88 dB at high speeds, while the cycloidal reducer generated stronger peaks, frequently climbing into the 92–95 dB range.

The limited observable wear during the conducted tests can be attributed in part to the use of a synthetic, polymer-compatible lubricant, as described in Sect. 3.3. The presence of the lubricant reduced direct asperity contact between the additively manufactured surfaces, thereby limiting abrasive and adhesive wear during operation. This effect is consistent with the experimental results, where lubrication led to a reduction in temperature rise and acoustic fluctuations without significantly altering the relative trends between the two reducer architectures.

To complement the thermal observations with an energy-based perspective, the electrical input power of the driving stepper motor was monitored under no-load conditions. The NEMA 23 motor was supplied at a constant 32 V DC, drawing approximately 1.5 A when running freely, corresponding to a baseline input of about 48 W associated with motor and driver losses. When coupled to the cycloidal reducer, the current increased progressively from 1.8 A at 100 rpm to 3.0 A at 600 rpm, yielding an additional power demand of 10–48 W across the tested speed range. In contrast, the harmonic reducer produced consistently higher current draw, reaching 3.9 A at 600 rpm, equivalent to an added consumption exceeding 75 W relative to the unloaded case. These differences indicate higher internal energy dissipation within the harmonic architecture, attributable to continuous elastic deformation of the flexural element and extended sliding contact at the tooth interfaces.

The contrasting thermal and acoustic responses of the two reducers can be traced directly to their fundamental kinematic principles. In the harmonic drive, the continuous elastic deformation of the flex spline combined with sliding tooth engagement concentrates frictional losses within a confined contact zone, leading to a progressive accumulation of heat as rotational speed increases. The moderate stiffness and layered

Table 2. Comparison of average thermal and acoustic results.

Parameter	Harmonic Drive	Cycloidal Drive
Average temperature at 600 rpm	Δ36 °C	Δ28.5 °C
Average sound level at 600 rpm	Δ89 dB	Δ93.5 dB
Temperature trend	Sharp rise with speed	Mild growth with speed
Acoustic character	Smooth and continuous tone	Pulsating, impact-driven noise

surface texture characteristic of the additively manufactured polymer components further intensify internal energy dissipation. By contrast, the cycloidal reducer exhibited lower thermal rise but produced higher acoustic emissions due to its discontinuous load transfer mechanism. The repeated transition of the cycloidal lobes between adjacent pins generates cyclic micro-impacts, which become increasingly pronounced at higher speeds and manifest as a pulsating noise pattern. While lubrication attenuated some high-frequency components, the dominant acoustic signature persisted, confirming that the elevated noise levels originate primarily from kinematic intermittency rather than surface friction alone.

5 Conclusions

The present study examined two additively manufactured transmission systems a harmonic reducer and a cycloidal mechanism under identical operating conditions, relying on thermal imaging and acoustic monitoring to reveal how each architecture manages friction, deformation, and internal load transfer. Because both units were produced from the same polymer on the same printer and with matching process parameters, the comparison isolated the mechanical response of the two designs rather than differences in material quality or fabrication.

Across all speed levels, the two reducers diverged in a consistent and predictable manner. The harmonic drive showed a marked tendency to accumulate heat, with temperatures rising to 36 °C at the highest speed stage, whereas the cycloidal mechanism stabilised around 28–29 °C under the same conditions. Acoustically, the situation reversed in a way that the harmonic unit produced a smoother and more uniform tone averaging around 88–90 dB at elevated speeds, while the cycloidal reducer exhibited stronger peaks and a distinctive pulsating pattern that frequently pushed sound levels into the 92–95 dB range. Although lubrication softened some of the extremes, the overall behaviour of both drives remained anchored in their internal kinematics. To highlight these differences more clearly, the main averaged results are summarised in Table 2 below.

Previous studies on harmonic and cycloidal transmissions, focused on metallic industrial implementations, report high mechanical efficiency, reduced backlash, and comparatively low acoustic emissions when manufacturing tolerances and surface finishes are carefully controlled [1, 3, 4]. More recent investigations on additively manufactured gear mechanisms highlight that polymer-based fabrication introduces fundamentally different tribological conditions, primarily due to increased surface roughness, anisotropic stiffness, and layer-induced compliance [23, 24]. In this context, the present study extends existing literature by providing a direct experimental comparison between harmonic and cycloidal reducers manufactured via fused deposition modelling under identical conditions. The results therefore bridge the gap between classical metal-based analyses and emerging research on 3D-printed transmissions.

Overall, the experimental results provide a coherent picture of how these two reducer types behave when produced through fused filament deposition. The harmonic drive dissipates a substantial fraction of its losses thermally, while the cycloidal arrangement expresses its internal dynamics acoustics acoustically. From a manufacturing sustainability and quality management perspective, the study demonstrates the usefulness of additive manufacturing as a low-waste, energy efficient platform for early-stage mechanical evaluation. By enabling rapid comparison of alternative gearbox concepts using accessible diagnostic tools, unfavourable tribological behaviours can be identified before committing to material and energy intensive production methods. While 3D-printed prototypes remain limited in terms of long-term wear resistance and mechanical efficiency, they offer a sustainable and effective means for kinematic validation, design refinement, and quality driven decision making in modern manufacturing workflows.

References

1. Sensinger, J.W., Lipsey, J.H.: Cycloid vs. harmonic drives for use in high ratio, single stage robotic transmissions. In: Proceedings of the IEEE International Conference on Robotics and Automation, pp. 4130–4135 (2012)
2. Xu, W.W.: A practical harmonic guideline for adjustable speed drive applications. IEEE Trans. Power Deliv. **7**, 427–433 (1992). https://doi.org/10.1109/61.108937
3. Buss, A., Trhoň, V.: Harmonic drive products in robotic applications. Appl. Mech. Mater. **613**, 112–116 (2014). https://doi.org/10.4028/www.scientific.net/AMM.613.112
4. Hu, Z., Liu, Y., Chen, Y., Xin, D.: Real-time wear monitoring in plastic gears using drive torque signal diagnostics. Results Eng. **28**, 107134 (2025). https://doi.org/10.1016/j.rineng.2025.107134
5. Li, X., et al.: Investigation on tooth surface wear of cycloid drives considering tooth profile modifications. Lubricants. **11**, 323 (2023). https://doi.org/10.3390/lubricants11080323
6. Xu, L., Li, S., Wang, W.: Sliding ratio for novel cycloidal gear drive. Proc. Inst. Mech. Eng. C J. Mech. Eng. Sci. **232**, 3954–3963 (2018). https://doi.org/10.1177/0954406217743272
7. Birleanu, C., et al.: Basalt vs. glass fiber-reinforced polymers: a statistical comparison of tribological performance under dry sliding conditions. J. Composites Sci. **9**, 444 (2025). https://doi.org/10.3390/jcs9080444
8. Yu, G., et al.: Examination on the wear process of polyformaldehyde gears under dry and lubricated conditions. Friction. **9**, 538–550 (2021). https://doi.org/10.1007/s40544-020-0362-7
9. Kamenev, S.V.: Building a conceptual model of the chain harmonic drive. J. Phys. Conf. Ser. **1333**, 052008 (2019). https://doi.org/10.1088/1742-6596/1333/5/052008
10. Liang, X., Luy, Y., Koval, D.O.: Investigation of input harmonic distortions of variable frequency drives. In: IEEE/IAS Industrial & Commercial Power Systems Technical Conference, pp. 1–11 (2007)
11. Chen, S.Y., Xin, H.B.: Speed torque measurement system design of micro harmonic drive. Appl. Mech. Mater. **155–156**, 455–458 (2012). https://doi.org/10.4028/www.scientific.net/AMM.155-156.455
12. Joseph, I.: Driveline fundamentals and lubrication. In: Chemistry and Technology of Lubricants, pp. 325–344. Springer, Dordrecht (2009)
13. Tichy, J.A., Meyer, D.M.: Review of solid mechanics in tribology. Int. J. Solids Struct. **37**, 391–400 (2000). https://doi.org/10.1016/S0020-7683(99)00101-8
14. Kumar, N., Kosse, V., Oloyede, A.: A novel testing system for a cycloidal drive. In: Proceedings of the International Conference on Intelligent Systems Design and Applications, pp. 503–507 (2012)
15. Qi, L., et al.: Design principle and numerical analysis for cycloidal drive considering clearance, deformation, and friction. Alex. Eng. J. **91**, 403–418 (2024). https://doi.org/10.1016/j.aej.2024.01.077

16. Nachimowicz, J., Rafałowski, S.: Modelling the meshing of cycloidal gears. Acta Mech. et Autom. **10**, 137–140 (2016). https://doi.org/10.1515/ama-2016-0022
17. Lin, W.-S., Shih, Y.-P., Lee, J.-J.: Design of a two-stage cycloidal gear reducer with tooth modifications. Mech. Mach. Theory. **79**, 184–197 (2014). https://doi.org/10.1016/j.mechmachtheory.2014.04.009
18. Wang, Z., Shen, X., Chen, X.: Mechanical characteristic analysis of micro gear mechanism. Integr. Ferroelectr. **161**, 112–122 (2015). https://doi.org/10.1080/10584587.2015.1036662
19. Żurowski, W., Olejarczyk, K., Zaręba, R.: Wear assessment of sliding sleeves in a single-stage cycloidal drive. Adv. Sci. Technol. Res. J. **13**, 239–245 (2019). https://doi.org/10.12913/22998624/114180
20. Prusa Polymers: Prusament PETG V0 – technical data sheet. Available online. Accessed 12 Sep 2025 (2023)
21. Jabłońska, M., Łastowska, O.: Enhancing of surface quality of FDM moulded materials through hybrid techniques. Materials. **17**, 4250 (2024). https://doi.org/10.3390/ma17174250
22. KANO Laboratories LLC: Super Lube multi-purpose synthetic grease with Syncolon – technical data sheet. Available online. Accessed 10 Nov 2025 (2024)
23. Birleanu, C., et al.: Tribomechanical analysis and performance optimization of sustainable basalt fiber polymer composites for engineering applications. Technologies. **13**, 249 (2025). https://doi.org/10.3390/technologies13060249
24. Birleanu, C., et al.: Experimental investigation of the tribological behaviors of carbon fiber reinforced polymer composites under boundary lubrication. Polymers. **14**, 3716 (2022). https://doi.org/10.3390/polym14183716

Assessment of Erosive Wear Performance of 3D Printed Industrial Cyclone Preheater

Rabia Sheraz[1], Rehan Khan[1](✉), Ayesha Maqsood[1], and Michał Wieczorowski[2]

[1] Department of Mechanical Engineering, College of Electrical and Mechanical Engineering, National University of Sciences and Technology, Islamabad, Pakistan
mrehan.khan@ceme.nust.edu.pk

[2] Faculty of Mechanical Engineering, Poznan University of Technology, Poznan, Poland

Abstract. Cyclone preheaters are widely employed in cement and process industries for heat recovery; however, internal erosion from particle-laden flows significantly reduces component life and increases maintenance costs. This study presents an integrated CFD–statistical–experimental approach to minimize erosion through geometric optimization. Five cyclone geometries were analyzed in ANSYS Fluent using a k–ε turbulence model coupled with a discrete-phase erosion model. A parametric study was conducted using main-effect analysis, regression, and ANOVA in Minitab to investigate the effects of inlet velocity, particle size, and flow rate on erosion, identifying key factors that influence it. The best-performing geometry (Design 2) was 3D-printed using PLA and tested in a closed sand–water loop, with observed wear patterns closely matching CFD predictions. Microscopic analysis confirmed mechanisms such as micro-cutting, ploughing, and crater formation in high-impact regions. Compared with the baseline, the optimized geometry reduced peak erosion rates by approximately 70%, demonstrating that targeted geometric modifications, particularly of inlet orientation and cone angle, offer a cost-effective strategy to enhance cyclone durability and operational efficiency.

Keywords: Erosive Wear · Cyclone Preheater · CFD Simulation

1 Introduction

Cyclone separators are widely used in cement production, mineral processing, and power generation for gas–solid separation. However, their operational lifespan is frequently limited by internal erosion caused by high-velocity particle impingement and abrasive wear. Erosion mechanisms, such as abrasive sliding, impact wear, and chemical attack, often act together, progressively damaging cyclone walls and reducing efficiency [1–3]. Previous studies have shown that erosion is strongly influenced by particle size and hardness [4], gas velocity and temperature [5] and geometric parameters including vortex-finder length, cone angle, and surface roughness [6, 7].

To mitigate this problem, several approaches have been explored. Material-based strategies, such as silicon-carbide and alumina linings [3, 8], metal–matrix composites and high-velocity oxy-fuel (HVOF) coatings using Ni60–WC–Cr_3C_2 have improved

M. Hryb et al. (Eds.): MANUFACTURING 2026, LNME, pp. 187–203, 2026.
https://doi.org/10.1007/978-3-032-23038-6_14

wear and corrosion resistance [9]. Geometric and flow-based improvements, including extended conical sections, hybrid square–circular designs, and adjustable vortex limiters, have been shown to alter flow distribution and reduce particle–wall collisions [10, 11]. Moreover, devices like laminarizers have been used to improve separation performance while minimizing localized erosion [12].

Although these advances have enhanced cyclone durability, most studies have focused on standard cyclone geometries or coating materials rather than optimizing preheater-specific designs under realistic flow conditions. Consequently, industrial cyclone preheaters still experience frequent shutdowns and high maintenance costs due to persistent erosion hotspots near the inlet, cone, and vortex finder [13, 14].

To address these gaps, this study proposes an integrated Computational Fluid Dynamics–statistical–experimental framework for erosion reduction in cyclone preheaters. Five geometric variants were modeled in ANSYS Fluent to evaluate the influence of cone angles, inlet configurations, and vortex-finder designs on erosion rates and particle impact behaviors. Statistical analysis in Minitab was used to quantify the relative effects of operating parameters, while experimental testing of the optimized 3D-printed model validated the numerical predictions. The outcomes of this research provide a systematic methodology for designing erosion-resistant cyclone preheaters, offering a practical, cost-effective solution to enhance reliability and energy efficiency in industrial systems.

2 Materials and Methods

Silica sand was selected as the erodent material to replicate realistic erosion conditions within the cyclone separator due to its angular particle shape, high hardness, and abrasive characteristics. Because of its high hardness, angular particle shape, and abrasive nature, it is frequently used in filtration and industrial erosion testing. The setup's particles, which had an average diameter of 300–600 μm, were appropriate for creating noticeable wear patterns on the surface.

This study used a combined CFD-based, statistical, and experimental approach in addition to material selection. ANSYS Fluent was used to conduct computational fluid dynamics (CFD) simulations to examine the gas–solid flow behavior, particle trajectories, and erosion rate distribution within the cyclone preheater.

Additionally, the relative effects of geometric and operating parameters on erosion rate were quantified by statistical analysis using Minitab. Lastly, to verify the simulation results under controlled flow circumstances, experimental testing was carried out on the optimized cyclone geometry created via 3D printing (Fig. 1).

Fig. 1. Microscopic Image of Silica Sand

3 Results and Discussion

3.1 CFD-Based Results

To examine the impact of cone angle, inlet configuration, and vortex-finder design on erosion behavior, five geometric variations of the cyclone preheater were modeled in SolidWorks. Erosion rates and flow characteristics for each design were evaluated in ANSYS Fluent under identical operating conditions. The CFD simulations carried out in ANSYS Fluent utilizing the integrated erosion prediction model produced the erosion rate (ER) values given in this study. Based on the impact angle, particle impact velocity, particle mass flow rate, and wall material characteristics, the model calculates erosion. The Discrete Phase Model (DPM), which tracks discrete solid particles and evaluates particle–wall interactions at each impact site, was used to determine the erosion rate. A semi-empirical erosion correlation was used to calculate the local erosion rate, and the integration of the local erosion contributions over the cyclone wall surfaces yielded the overall erosion rate. Table 1 illustrates all five cyclone preheater designs used in this study, while Table 2 presents the corresponding CFD-based results for each configuration. The optimized cyclone design (Design 2) is evaluated experimentally and numerically in this section. Thus, only the comprehensive CFD data and erosion outlines of the final, chosen configuration are shown here, followed by the associated experimental validation (Fig. 2).

Table 1. Geometric Configurations of Cyclone Preheater Designs (Designs 1–5)

Design Number	Key Geometric Feature	CAD Image
Design 1	Baseline industry-standard geometry	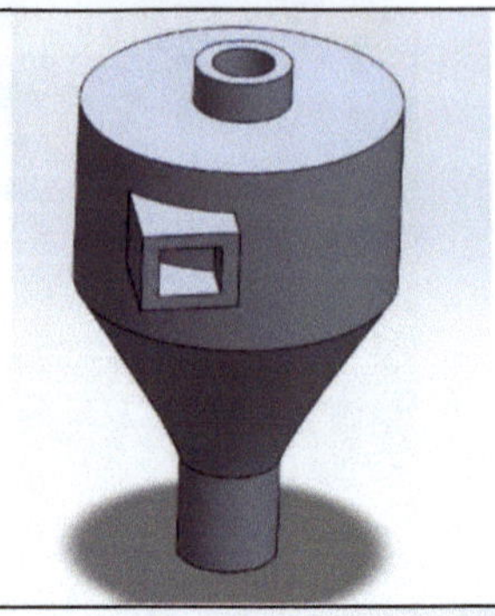
Design 2 (Selected)	Optimized cone angle (~30°) + refined inlet	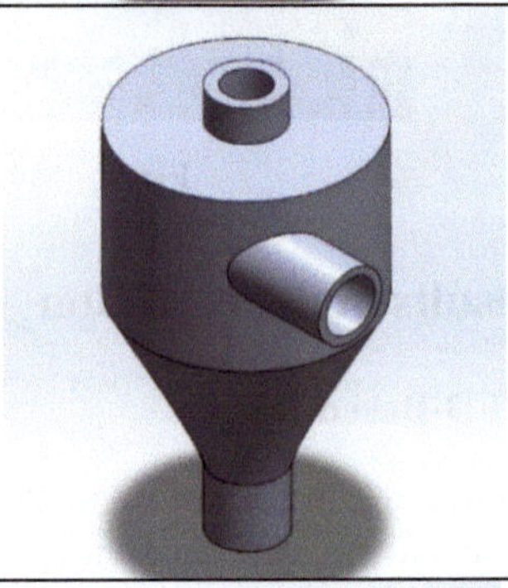
Design 3	Narrow cone (~15°), elongated body	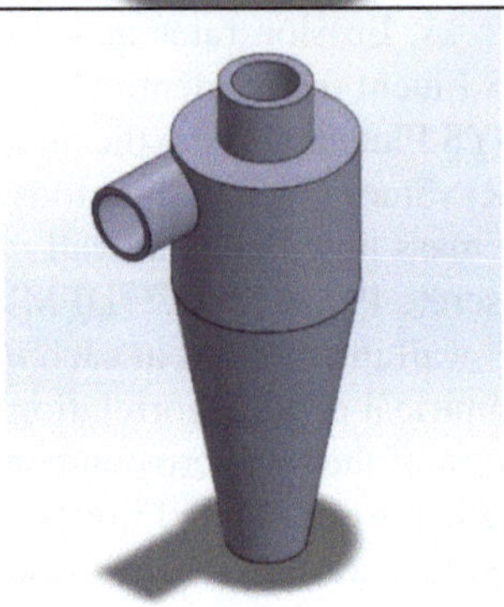

(*continued*)

Table 1. (*continued*)

Design 4	Tangential inlet with improved vortex finder	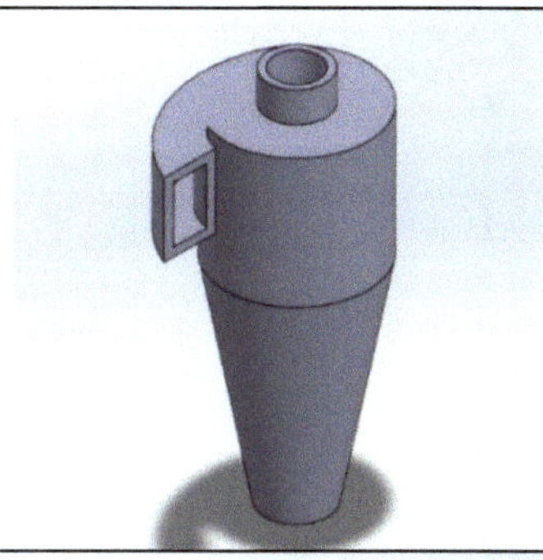
Design 5	Hybrid features from design (1–4)	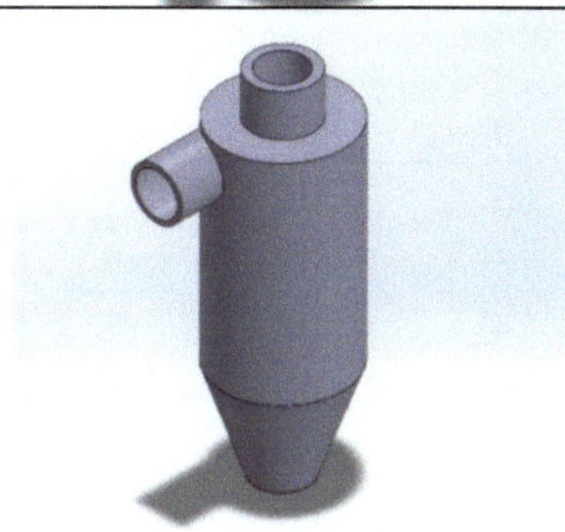

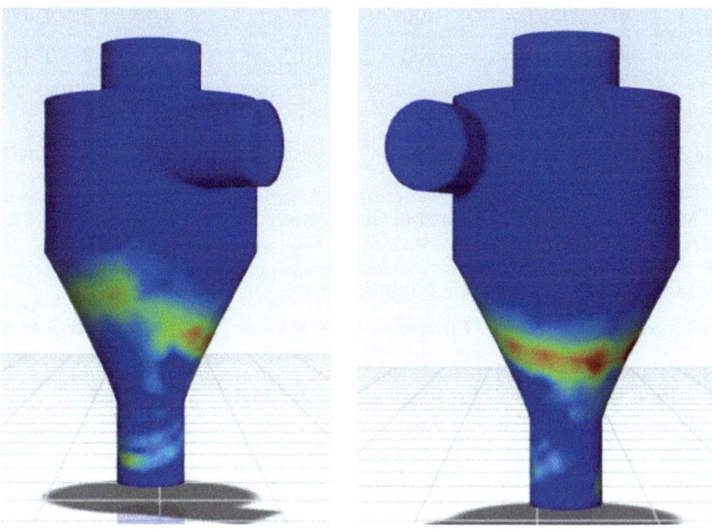

Fig. 2. Design 2 Erosion Pattern Case 1.

Table 2. Summary of CFD Results

Case	Design	Velocity (m s^{-1})	Size (μm)	Flow rate (kg s^{-1})	ER (kg m^{-2} s^{-1})
1	1	10	200	0.5	8.18E-03
2	1	20	300	1	1.02E-03
3	1	30	400	1.5	8.02E-03
4	1	40	500	2	1.09E-01
1	2	10	200	1.5	2.62E-05
2	2	20	300	2	2.75E-05
3	2	30	400	0.5	7.30E-06
4	2	40	500	1	9.80E-05
1	3	10	200	2	8.16E-05
2	3	20	300	1.5	1.07E-02
3	3	30	400	1	1.02E-02
4	3	40	500	0.5	2.15E-01
1	4	10	200	1	1.19E-02
2	4	20	300	0.5	1.09E-04
3	4	30	400	2	5.76E-03
4	4	40	500	1.5	6.15E-03
1	5	10	200	1	4.25E-05
2	5	20	300	0.5	2.34E-04
3	5	30	400	2	2.60E-03
4	5	40	500	1.5	2.37E-03

Table 3. Performance Comparison of Cyclone Designs Based on CFD Analysis

Design	Max Erosion Rate (kg m^{-2} s^{-1})	Notable Observation	Final Status
1	1.09×10^{-1}	High inlet/cone erosion	Rejected
2	9.8×10^{-5}	Lowest erosion; smooth flow	**Selected & 3-D printed**
3	2.15×10^{-1}	Severe cone wear at high velocity	Rejected
4	8.5×10^{-4}	Moderate erosion	Rejected
5	2.37×10^{-3}	Localized wear near the inlet	Rejected

Table 3 summarizes the maximum erosion rate of each design. The configuration with the lowest predicted erosion (Design 2) was selected for prototype fabrication and experimental validation.

Low flow rates promoted unstable particle deposition, as indicated by main-effect analysis, which showed that erosion was highest at 0.5 kg s^{-1} and decreased significantly at 1–1.5 kg s^{-1}, with a slight rise at 2 kg s^{-1}Similar trends apply to particle size: erosion is minimal for 200–400 μm but increases dramatically at 500 μm due to the higher impact energy that larger particles provide. With little wear at 10–30 m s^{-1} and a sharp surge at 40 m s^{-1} because of increased kinetic energy, velocity is the primary contributor. Design 2 exhibits the best control and the lowest rate of erosion among the five geometries, but Design 3 has the worst wear. Designs 1, 4, and 5 perform moderately but remain inferior to Design 2 (Fig. 3).

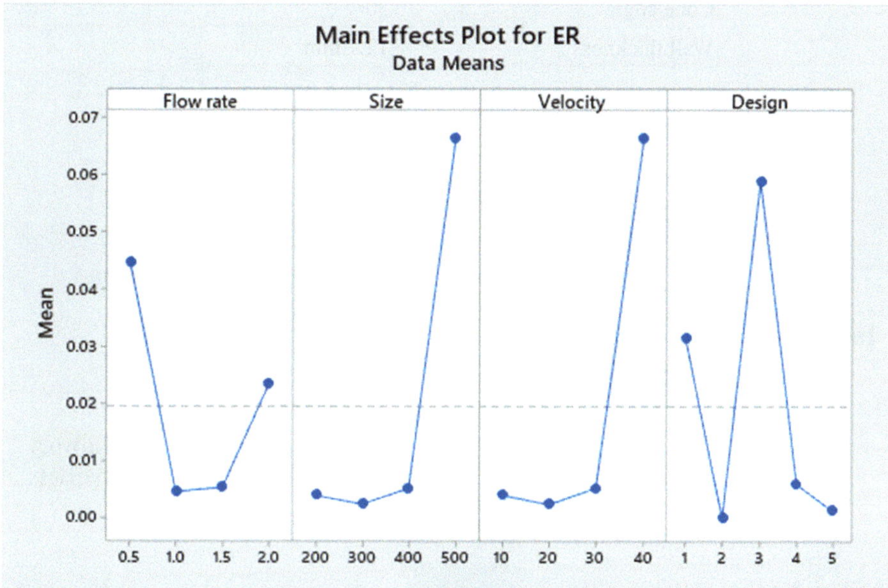

Fig. 3. Main Effect Plot.

3.2 Prototype Preparation

Design 2 had the lowest erosion rate among all the designs analyzed. Based on the findings of the simulation study, it was selected for experimental testing. The selected design was made using 3D printing technology to allow for experimental validation and guarantee accurate geometrical reproduction. The geometric dimensions of the selected and 3D printed Cyclone Preheater Design are shown in Table 4.

Table 4. Geometric Dimensions of Selected Cyclone Preheater Design

Parameter	Value (mm)
Cyclone body diameter	203.2 mm
Total cyclone height	330.2 mm
Inlet diameter	50.8 mm
Inlet length	63.5 mm
Upper outlet diameter	50.8 mm
Upper outlet length	50.8 mm
Lower outlet diameter	38.1 mm
Lower outlet length	76.2 mm
Cone angle	30°
Wall thickness	12.7 mm

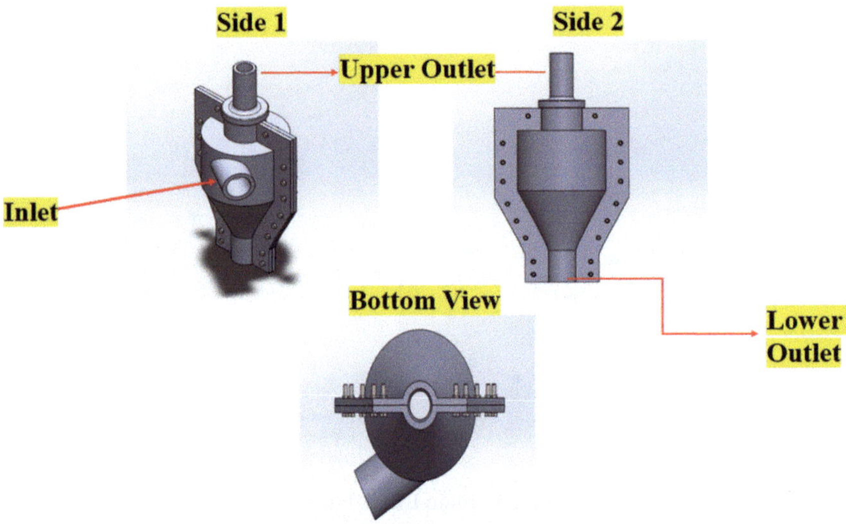

Fig. 4. Preparation of Geometry for 3D Printing.

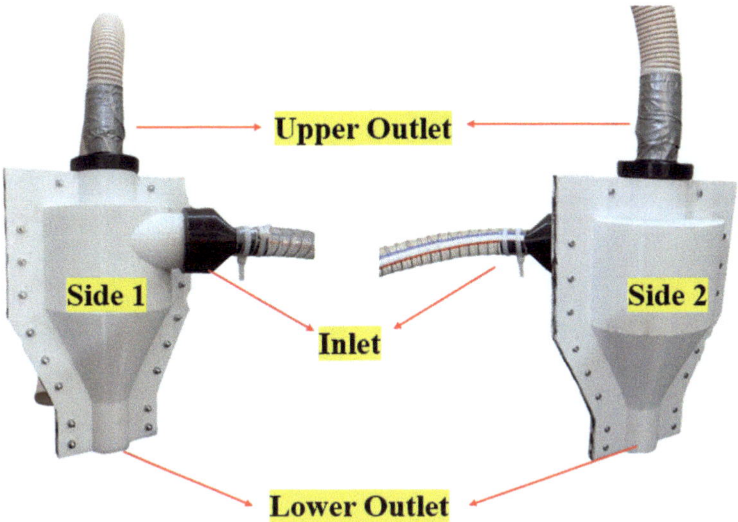

Fig. 5. 3D Printed Cyclone Model.

The 3D-printed model was fabricated using durable polymer material (PLA) suitable for handling sand-water mixtures under flow conditions. Material selection ensured sufficient strength while allowing observation of erosion effects post-experimentation (Figs. 4 and 5).

The fabricated cyclone prototype was integrated into a closed-loop experimental setup designed to replicate industrial flow conditions. The setup included:

- A pump to circulate the sand-water mixture.
- An inlet connection feeding the mixture into the cyclone.
- Two outlets: the upper outlet returning clear water into the drum and the lower outlet discharging sand-laden water, following the principle of particle separation (Fig. 6).

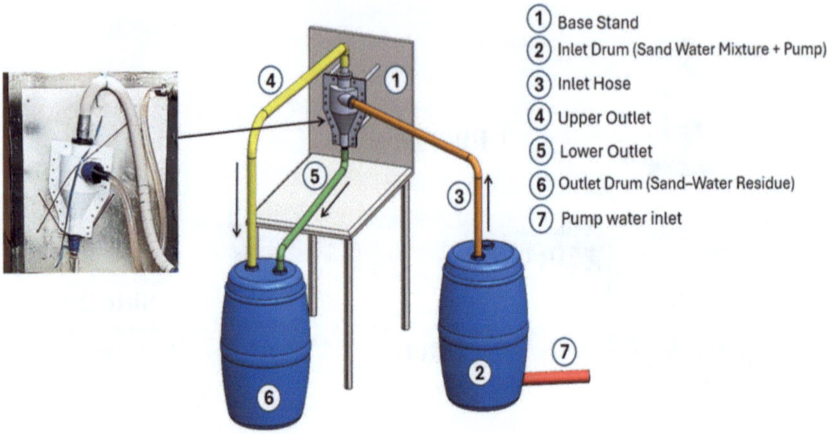

Fig. 6. Experimental Setup.

The setup effectively simulated real operational conditions, allowing for a continuous flow over an extended period to study the wear behavior of the cyclone model.

3.3 Experiment-Based Results

Following continuous operation for 10 h, the 3D-printed cyclone model was inspected for signs of surface degradation, the 3D-printed cyclone separator model was thoroughly inspected for indications of material degradation. Distinct wear marks and surface erosion were observed, particularly in regions subjected to intense particle impingement, closely aligning with CFD-predicted erosion zones, especially in areas exposed to intense particle impingement, closely aligning with the high-erosion zones predicted in the CFD analysis. In some zones, visible surface degradation and material loss confirmed the erosion-prone areas identified during simulation. The wear patterns aligned well with the expected high-velocity and high-impact regions (Figs. 7 and 8).

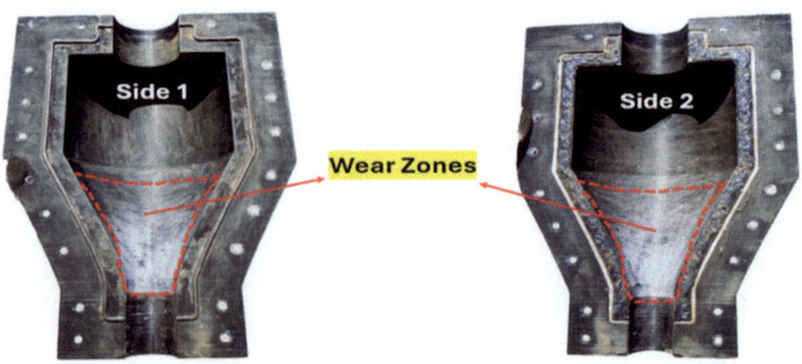

Fig. 7. Demonstration of wear zones.

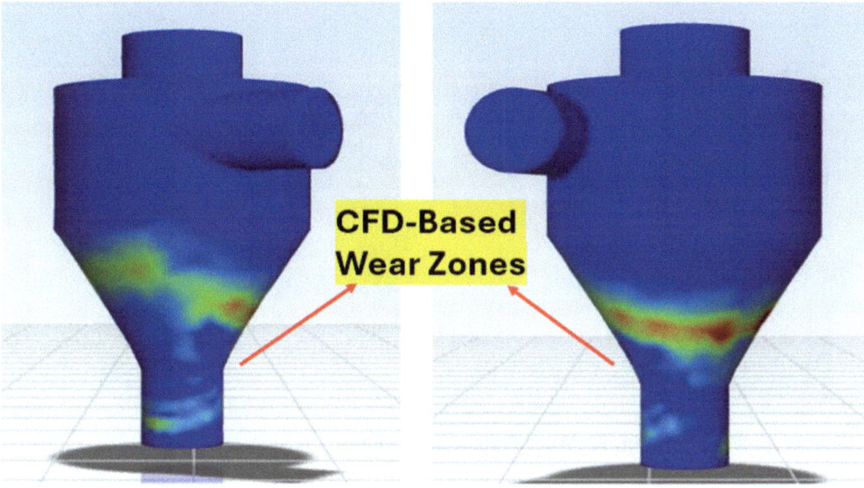

Fig. 8. CFD Results for Wear Zones.

The experimental results obtained from the test setup show a strong agreement with the simulation outcomes. Both analyses indicate similar erosion trends and material loss patterns under comparable operating conditions. The minor deviations observed can be attributed to practical factors such as flow irregularities, measurement uncertainties, and surface roughness variations, which are difficult to replicate precisely in simulations. Overall, the close correlation between experimental and simulated data validates the accuracy of the numerical model and confirms that the adopted simulation parameters effectively represent the real system behavior.

The wear distribution along the cyclone preheater wall exhibits a clear gradient from the top to the bottom section, as shown in Figs. 9 and 10. The conical region demonstrates the highest degree of wear, primarily due to the intensified spiral particle motion and elevated centrifugal forces, which promote frequent high-angle particle–wall impacts. These interactions lead to dominant abrasive wear mechanisms such as micro-cutting, ploughing, and surface flaking. The transition zone between the cylindrical and conical sections experiences moderate wear, where reduced particle velocity and impact frequency result in groove formation, moderate pitting, and initial delamination. Conversely, the upper cylindrical section is subjected to minimal wear, attributed to its relatively stable flow pattern and lower particle kinetic energy, which primarily cause minor surface polishing and scattered micro-scratches.

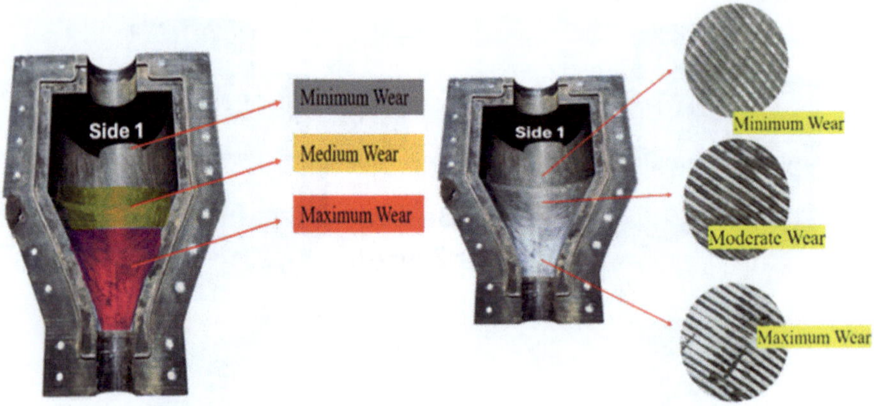

Fig. 9. Wear Pattern Distribution on Side 1.

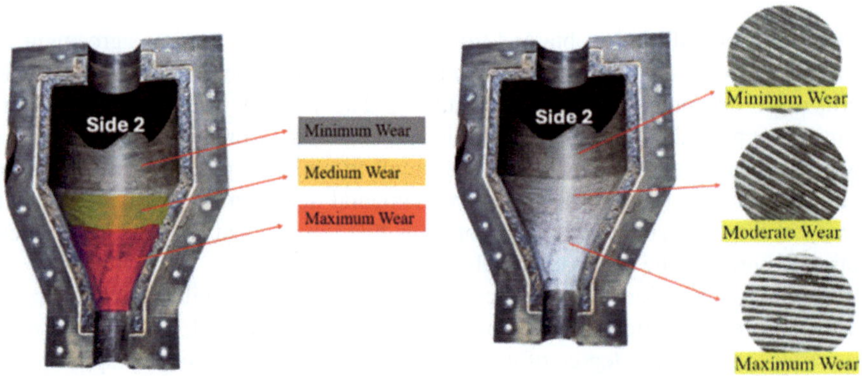

Fig. 10. Wear Pattern Distribution on Side 2.

Microscopic examination of the cyclone separator's inner surface revealed distinct deformation mechanisms corresponding to varying wear intensities across different regions. The maximum wear zone exhibited severe damage characterized by chipping, micro-cutting, ploughing, and erosive fissures, indicating extensive plastic deformation caused by high-velocity particle impacts. In the moderate wear zone, surface degradation was less pronounced, dominated by fractured and chipped regions suggestive of brittle failure under repeated particle collisions. Conversely, the minimum wear zone showed negligible erosion, with embedded anthracite particles observed on the surface, implying minimal particle impingement and low-velocity flow conditions that limited material removal.

(a)

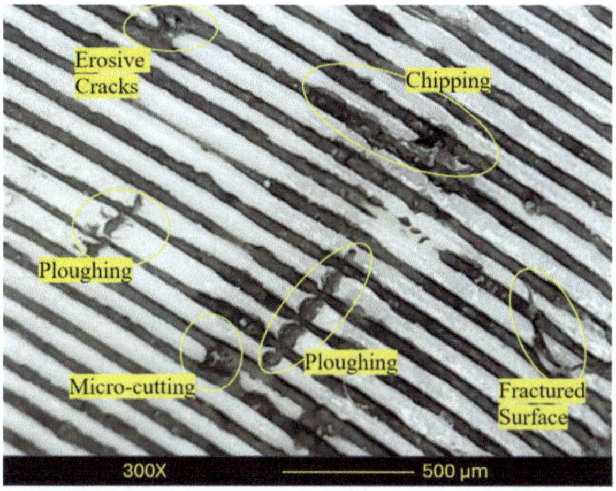

(b)

(c)

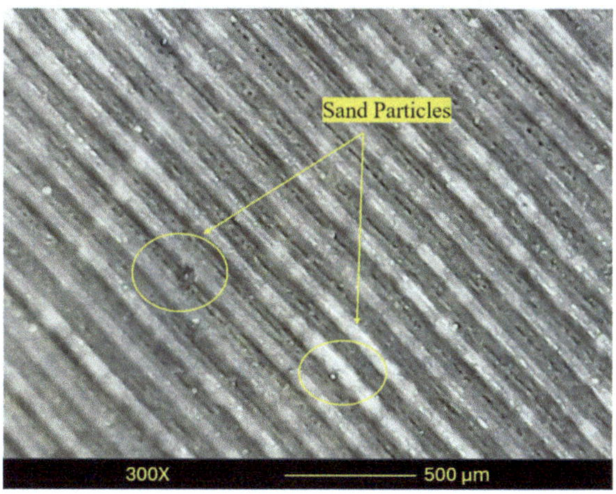

Fig. 11. (a) Maximum wear, (b) Moderate wear, (c) Minimum wear.

Microscopic examination of the internal surface of Side 2 revealed distinct wear characteristics corresponding to varying erosion intensities. The maximum wear region exhibited severe material degradation marked by deep ploughing grooves, fractured surfaces, and crater-like pits formed by high-velocity particle impacts, causing plastic deformation and localized fatigue failure. In the moderate wear region, the surface showed comparatively lower damage, dominated by fractured areas and shallow craters indicative of reduced impact energy and particle interaction. The minimum wear region, however, remained largely unaffected, displaying only minor surface cracks likely caused by occasional low-energy abrasive contacts.

Microscopic analysis revealed that the cyclone's inner surface experienced severe wear in high-impact zones due to intense particle erosion, showing features like craters, ploughing, and micro-cutting. In contrast, the moderate zone showed limited surface damage, while the minimal wear zone exhibited only slight cracking with negligible material loss. Based on a visual examination of the worn cyclone surfaces, the erosion severity seen in Figs. 9, 10, 11, and 12 was categorized qualitatively. The degree of apparent material removal, surface discoloration, roughness, and erosion mark density were taken into consideration when defining the classification criteria.

Minimum wear areas have smooth surfaces, very little material removal, and little to no erosion. Partial material loss, obvious scratch patterns, and surface roughening were all present in moderately worn areas. Strong surface deterioration, a high density of scratches and grooves, obvious material loss, and noticeable discoloration—all of which point to severe particle impingement—were used to identify maximum wear zones. A consistent comparison of erosion severity across various cyclone configurations was made possible by this qualitative classification.

It should be noted that the erosion assessment presented in this study is primarily based on qualitative visual inspection of wear patterns on the cyclone surfaces. Detailed quantitative measures, such as mass loss calculation and surface topography analysis, were not carried out due to experimental and time constraints during the study's execution. However, the observed qualitative erosion patterns consistently agree with the erosion zones predicted by CFD, offering trustworthy information about the relative severity of erosion across various cyclone configurations. It is advised that future research should include quantitative surface characterization to improve the precision of erosion evaluation.

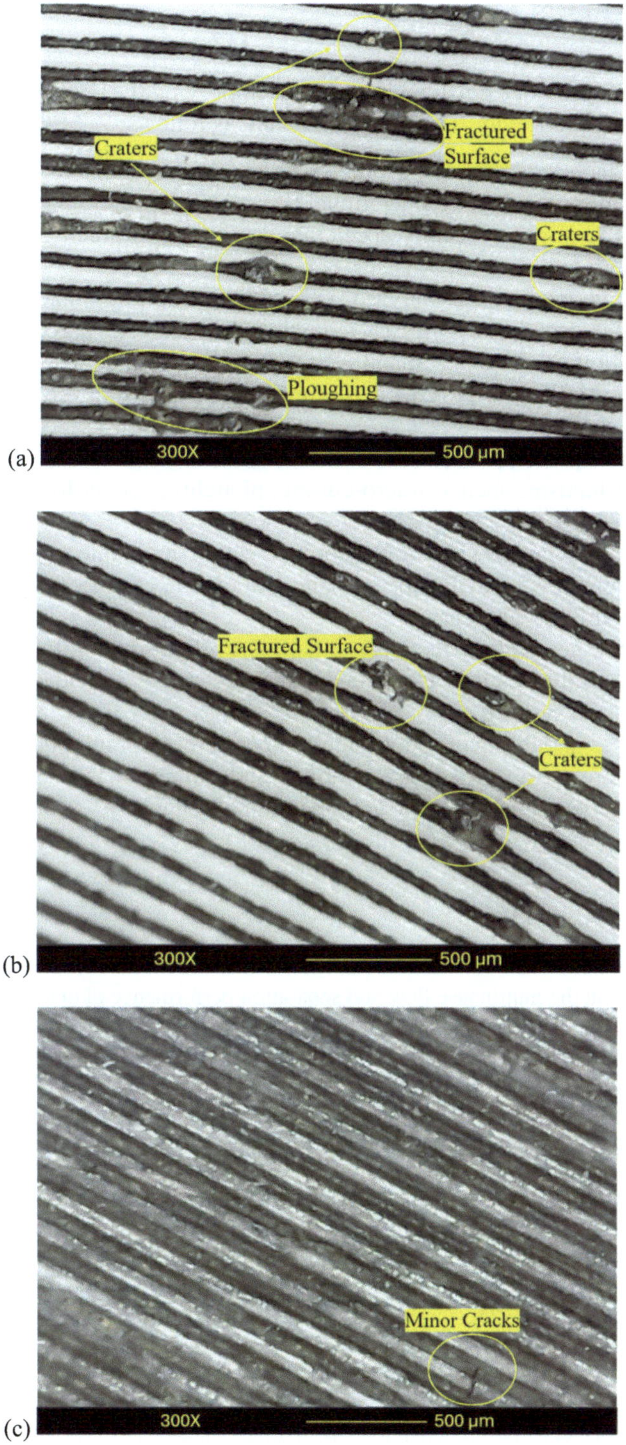

Fig. 12. (a) Maximum wear, (b) Moderate wear, (c) Minimum wear.

4 Conclusions

The investigation provides practical insights for improving the design and operation of industrial cyclone preheaters. The key findings are as follows:

- Erosion rate increases sharply beyond 35–40 m s^{-1}, indicating that controlling inlet velocity is the most effective operational parameter for minimizing wear. Particles larger than 400 μm significantly accelerate erosion due to their higher momentum and impact energy.
- Among the five geometries analyzed, Design 2, with an optimized cone angle and tangential inlet, achieved over a 70% reduction in erosion compared with the conventional configuration. Regression and ANOVA confirmed that cyclone geometry acts as a "force multiplier," amplifying or diminishing the effects of velocity and flow-rate variations.
- The 3D-printed prototype reproduced CFD-predicted erosion hotspots and validated observed mechanisms such as micro-cutting, ploughing, crater formation, and crack propagation.
- Maximum wear occurred in the lower conical section, moderate wear in the transition zone, and minimal wear in the upper cylindrical region, consistent with the predicted turbulence and flow behavior.
- The optimized geometry effectively enhances cyclone durability, reduces maintenance costs, minimizes downtime, and improves energy efficiency in cement and related industrial systems.
- Further studies should incorporate high-temperature and multiphase conditions, as well as explore advanced surface coatings and data-driven design approaches to achieve greater erosion resistance and operational reliability.

References

1. Chu, K.W., Kuang, S.B., Yu, A.B., Vince, A., Barnett, G.D., Barnett, P.J.: Prediction of wear and its effect on the multiphase flow and separation performance of dense medium cyclone. Miner. Eng. **56**, 91–101 (2014). https://doi.org/10.1016/J.MINENG.2013.10.029
2. Han, C., Hu, Y., Li, W., Bie, Q.: Gas-solid separation performance and structure optimization in 3D printed guide vane cyclone separator. Adv. Powder Technol. **33**(11), 103815 (2022). https://doi.org/10.1016/J.APT.2022.103815
3. Medvedovski, E.: Advanced ceramics and coatings for wear and corrosion related applications in modern high-efficient coal production and processing: a technical review. Ceram. Int. **50**(11), 19447–19487 (2024). https://doi.org/10.1016/J.CERAMINT.2024.03.187
4. Gao, Z., et al.: Peculiarities of particle motion inside cyclone separator by using LES-DRW model. Chem. Eng. Res. Des. **183**, 512–524 (2022). https://doi.org/10.1016/J.CHERD.2022.05.053
5. Brar, L.S., Rahmani, F.: The impact of operating temperatures on the fluctuating flow field and precessing vortex core in cyclone separator using large-eddy simulations. Phys. Fluids. **36**(3) (2024). https://doi.org/10.1063/5.0195382
6. Dehdarinejad, E., Bayareh, M.: Impact of non-uniform surface roughness on the erosion rate and performance of a cyclone separator. Chem. Eng. Sci. **249**, 117351 (2022). https://doi.org/10.1016/J.CES.2021.117351

7. Dehdarinejad, E., Bayareh, M., Parvaz, F., Hosseini, S.H., Ahmadi, G.: Performance analysis of a gas cyclone with a converging-diverging vortex finder. Chem. Eng. Res. Des. **193**, 587–599 (2023). https://doi.org/10.1016/J.CHERD.2023.04.012
8. Khorasanimohammadrezaei, M., Aminizadeh, N.: The impact of a ceramic wear liner on the separation efficiency of a particular cyclone dust collector. Procedia Eng. **157**, 414–421 (2016). https://doi.org/10.1016/J.PROENG.2016.08.384
9. Yang, M., Song, P., Kong, D., Huang, T., Jin, Q.: Wear and corrosion properties of HVOF sprayed WC-Cr3C2 composite coating for application in polysilicon cyclone separator. J. Mater. Res. Technol. **29**, 78–89 (2024). https://doi.org/10.1016/J.JMRT.2024.01.114
10. Parvaz, F., Hosseini, S.H., Elsayed, K., Ahmadi, G.: Numerical investigation of effects of inner cone on flow field, performance and erosion rate of cyclone separators. Sep. Purif. Technol. **201**, 223–237 (2018). https://doi.org/10.1016/J.SEPPUR.2018.03.001
11. Haake, J., Oggian, T., Utzig, J., Rosa, L.M., Meier, H.F.: Investigation of the pressure drop increase in a square free-vortex cyclonic separator operating at low particle concentration. Powder Technol. **374**, 95–105 (2020). https://doi.org/10.1016/J.POWTEC.2020.07.008
12. Huang, A.N., et al.: Influence of a laminarizer at the inlet on the classification performance of a cyclone separator. Sep. Purif. Technol. **174**, 408–416 (2017). https://doi.org/10.1016/J.SEPPUR.2016.09.053
13. Mirzaei, M., et al.: Investigation of erosion in an industrial cyclone preheater by CFD simulations. Powder Technol. **421** (2023). https://doi.org/10.1016/j.powtec.2023.118424
14. Zhang, L., Fan, J., Zhang, P., Gao, F., Chen, G., Li, J.: Effect of local erosion on the flow field and separation performance of the cyclone separator. Powder Technol. **413**, 118007 (2023). https://doi.org/10.1016/J.POWTEC.2022.118007

Digitalization and Automation of Quality Control Systems in the Factory of the Future

An Expert Decision Quality Assessment Model Based on Grey Relational Analysis and Kappa Coefficient

Liangyan Tao[1], Zhao Yang[1], Magdalena Hryb[2(✉)], and Agnieszka Kujawińska[2]

[1] College of Economics and Management, Nanjing University of Aeronautics and Astronautics, Nanjing, People's Republic of China
[2] Faculty of Mechanical Engineering, Poznan University of Technology, Poznan, Poland
magdalena.hryb@put.poznan.pl

Abstract. This study addresses the challenges of subjectivity and inconsistency in expert-based wood quality assessment by proposing a novel hybrid model. The model integrates Grey Relational Analysis (GRA) and the Kappa coefficient to provide a systematic, two-dimensional evaluation, simultaneously quantifying expert decision accuracy and consistency. The framework is applied to industrial data from ten experts assessing 28 wood quality attributes. Categorical grades are processed via numerical encoding and a grey possibility function. A key analysis compares the reliability of GRA and the weighted Kappa coefficient under small-sample conditions (10–20 samples). Results show that while both methods produce a coherent expert ranking, GRA exhibits superior stability and lower variability with limited data, suggesting greater robustness in such scenarios. The model also identifies specific weaknesses in the evaluation process, particularly for borderline qualification grades, providing actionable insights for training. The proposed tool offers a practical solution for expert evaluation in data-scarce environments. Its main limitation is the dependence on a subjective expert reference standard, indicating a need for validation with objective benchmarks and larger datasets.

Keywords: Expert Decision-Making · Decision Quality Evaluation · Grey Relational Analysis · Kappa Coefficient

1 Introduction

Wood quality assessment serves as the foundation of the wood production industry. Without guaranteed quality, the industry's overall level of development remains constrained. In practice, however, expert-based wood quality evaluations are highly subjective, and the accuracy of these judgments is influenced by factors such as the expert's knowledge limitations, personal preferences, and perceptual biases resulting from long-term exposure to specific wood characteristics. Historically, wood quality assessment has relied heavily on empirical judgment, often lacking systematic and scientific analytical rigor.

M. Hryb et al. (Eds.): MANUFACTURING 2026, LNME, pp. 207–219, 2026.
https://doi.org/10.1007/978-3-032-23038-6_15

Unlike objective, metric-driven methods in materials engineering - such as brushing studies quantifying defect removal [1–3] - wood assessment remains reliant on subjective judgment. This study introduces a dual-dimensional model to bring systematic, quantitative evaluation to expert-based wood quality assessment.

Jozsa and Middleton [4] revealed the causes of key wood quality attributes and their practical effects on processing and utilization through anatomical, physiological, and forestry research. Beyond wood science, expert evaluation serves as an indispensable tool for decision support, quality control, and risk management across numerous industries. Yet this approach is not without drawbacks: experts are susceptible to personal biases and cognitive limitations during evaluation, leading to subjective results. Kozierkiewicz [5] proposed a framework for assessing the quality of final decisions, based entirely on analyzing inputs such as hypothetical representations of expert opinions. Dror and Charlton [6] observed that under identical conditions, nearly two-thirds of experts made inconsistent decisions when reviewing past judgments. They categorized errors into three types: human errors, instrumental and technical errors, and methodological errors. Their work particularly highlights the third category - errors rooted in cognitive mechanisms - which can arise even when experts are qualified and equipment is functioning properly. They note that the dynamic and active nature of human cognition can lead to information-processing biases, to which even experts are not immune. Geva and Maytal [7] proposed a data-driven method called REQ (Ranking Expert decision makers' unobserved decision Quality) for ranking experts based on historical data in settings where decision quality cannot be directly observed - such as when true outcomes are unavailable. This method holds significant value in fields like medical diagnosis and tax auditing, which heavily rely on expert judgment but face verification challenges.

Grey Relational Analysis (GRA) [8, 9] is an active branch of grey system theory. Its basic idea is to judge the closeness of the relationship between different sequences based on the geometric shape of sequence curves. The core approach involves converting discrete behavioral observations of system factors into piecewise continuous polyline segments through linear interpolation, and then constructing a model to measure the degree of correlation based on the geometric characteristics of these polylines. GRA has been widely applied in various fields, such as the evaluation of socio-economic performance of countries [10], transport safety analysis [11], and UAV Combat Effectiveness Assessment [12]. The Kappa coefficient is a statistical measure used to evaluate agreement between two observers assessing the same objects, or between repeated assessments by the same observer [13, 14]. It quantifies the difference between observed agreement and agreement expected by chance, and has been widely adopted in practice [15]. Depending on the data type, commonly used variants include the simple Kappa coefficient for nominal categories and the weighted Kappa coefficient for ordinal data [16].

This study aims to develop a systematic method for evaluating expert decision quality in wood quality assessment and to rank experts based on their decision-making capabilities. We propose a dual-dimensional evaluation model that integrates Grey Relational Analysis (GRA) and the Kappa coefficient to quantify both the accuracy and stability of expert decisions.

The remainder of the paper is organized as follows. Section 2 describes the problem of expert decision quality assessment. Section 3 details the model construction, divided into two subsections: Sect. 3.1 presents the expert decision quality assessment model based on grey relational analysis, and Sect. 3.2 introduces the model based on the Kappa coefficient. Section 4 then provides a comparative analysis of GRA and the Kappa coefficient in the context of expert decision quality assessment. Finally, Sect. 5 concludes the study.

2 Problem Description

The research data used in this paper are sourced from a Polish floorboard manufacturer (more information on the authors' research into floorboard - [17, 18]). The company employs quality inspectors to evaluate the quality of its wood materials. The problem addressed in this study is the ranking of evaluators' decision accuracy - defined here as the quality of their assessments. Specifically, ten evaluators assessed the same batch of wood based on 28 characteristic attributes, assigning an overall quality grade according to a comprehensive evaluation of these attributes. The most experienced evaluator was designated as the reference (dec), and the remaining nine formed the comparison group. The structure of the dataset is presented in Table 1.

Table 1. Schematic of the Evaluator Decision Dataset. Own study.

Samples	c_1	c_2	c_3	c_4	c_5	...	c_{26}	c_{27}	c_{28}	dec
1	0	1	0	0	0		0	0	0	A
2	0	0	0	1	0		1	0	0	C
3	1	0	0	0	0		1	0	1	AB
4	0	1	0	0	0		1	0	1	B
5	0	0	1	0	0		0	1	0	NOK
...	0	0	0	1	0		1	0	0	CD

In Table 1, c_1 to c_{28} represent the quality attributes of wood, and the last column "dec" represents the wood grade (correct decision), where Grade A is the highest quality and NOK indicates unqualified wood.

Based on the problem description and data characteristics outlined above, this chapter employs a combined methodology integrating Grey Relational Analysis (GRA) and the Kappa coefficient to address the research objective. The grey relational analysis is applied to evaluate the performance of the control group against the reference evaluator along two dimensions: first, by measuring the grey relational degree between each evaluator and the reference evaluator in terms of their final decision outcomes; and second, by assessing the judgment consistency between each evaluator and the reference evaluator for each individual characteristic attribute. The Kappa coefficient is then utilized to specifically verify the level of agreement between each evaluator and the reference evaluator regarding the final assigned quality grades.

All data used in this study are complete, with no missing values across the 28 attributes of each decision sequence or in the final grading decisions. Consequently, no data imputation or interpolation methods were required.

3 Model Construction

3.1 Expert Decision Quality Assessment Model Based on Grey Relational Analysis

Since the classic grey relational model mainly processes numerical data, while the research object of this paper is ABC grade data, this paper adopts two processing methods to convert the decision results of each evaluator into numerical codes, namely: the method based on conventional numerical encoding and the conversion using grey possibility function. Conventional numerical encoding converts decision grades such as A, B, and C into calculable numbers like 1, 2, and 3 in sequence. This approach not only converts non-calculable characters into quantifiable numbers but also effectively quantifies the penalty degree for decision errors. For example, misclassifying wood of Grade A as Grade B or C will result in different decision error penalties according to the spans of $1 \rightarrow 2$ and $1 \rightarrow 3$ in the calculation. The larger the misclassification span, the higher the penalty degree.

Regarding the use of the grey possibility method for analysis, we first introduce the concept of the grey possibility degree. It is used to describe the degree to which an element belongs to a certain grey category. The possibility function maps each element in the domain to a real number within the interval [0, 1], which represents the possibility degree of the element corresponding to the grey category. A possibility degree closer to 1 indicates a higher degree of the element belonging to the grey category; conversely, the degree is lower. According to the standard gradient of wood quality grades, we define the possibility degree vector for each label in Table 2.

We take the reference evaluator and Evaluator 1 as examples to illustrate: according to Table 2, we first construct the decision matrices of the reference evaluator and Evaluator 1 respectively.

$$X_{Ref} = \left| [1,0,0,0,0]\ [1,0,0,0,0] \vdots \left[0,0,0,0,1\right] \right|_{30x5} \tag{1}$$

$$X_{E1} = \left| [1,0,0,0,0]\ [1,0,0,0,0] \vdots \left[0,0,0,0,1\right] \right|_{30x5} \tag{2}$$

Where the subscripts Ref and E1 represent the reference expert and Evaluator 1 to be assessed, respectively. Then, the difference matrix is calculated.

$$\Delta 1 = X_{Ref} - X_{E1} = \left| [0,0,0,0,0]\ [0,0,0,0,0] \vdots \left[0,0,0,0,0\right] \right|_{30x5} \tag{3}$$

According to the formula for calculating the Euclidean distance between two matrices, the distance between the matrices shown in (1) and (2) is:

$$d_F^1\left(X_{Ref}; X_{E1}\right) = |\Delta 1|_F \tag{4}$$

Table 2. Possibility Degree Conversion. Own study.

Label	Dimension 1 (A)	Dimension 2 (B)	Dimension 3 (C)	Dimension 4 (D)	Dimension 5 (NOK)	Explanation
A	1.0	0.0	0.0	0.0	0.0	Typical Grade A
AB	0.5	0.5	0.0	0.0	0.0	Between A and B
B	0.0	1.0	0.0	0.0	0.0	Typical Grade B
BC	0.0	0.5	0.5	0.0	0.0	Between B and C
C	0.0	0.0	1.0	0.0	0.0	Typical Grade C
CD	0.0	0.0	0.5	0.5	0.0	Between C and D
D	0.0	0.0	0.0	1.0	0.0	Typical Grade D
D0	0.0	0.0	0.0	0.8	0.2	Between D and Unqualified
0D	0.0	0.0	0.0	0.2	0.8	Between Unqualified and D
NOK	0.0	0.0	0.0	0.0	1.0	Unqualified
X	0.2	0.2	0.2	0.2	0.2	Unknown Grade

where $| \cdot |F$ is the Frobenius norm, a commonly used matrix norm. Next, we calculate the distances between the reference evaluator and the other 8 evaluators following the above steps to obtain the distance vector:

$$\left(d_F^1, d_F^2, d_F^3, d_F^4, d_F^5, d_F^6, d_F^7, d_F^8, d_F^9\right) \tag{5}$$

The maximum and minimum distances are obtained from Eq. (5), denoted as Δmax and Δmin respectively. Then, the grey relational degree between the *i-th* evaluator and the reference evaluator is defined as follows:

$$\gamma_i = \frac{\Delta_{min} - \rho \cdot \Delta_{max}}{d_F^i + \rho \cdot \Delta_{max}} \tag{6}$$

where ξ is the distinguishing coefficient, generally set to 0.5.

The grey relational degree rankings obtained by the two methods are shown in Fig. 1.

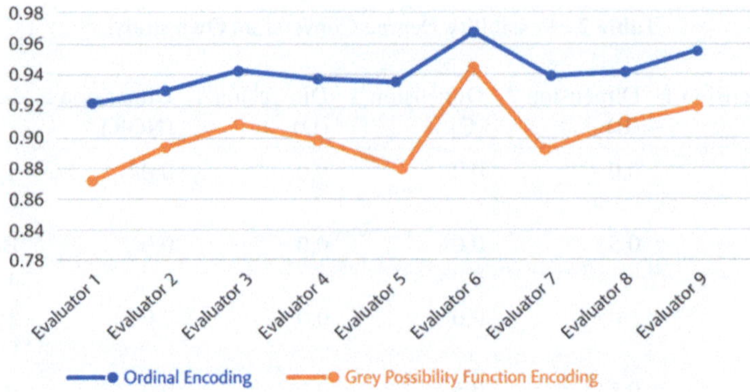

Fig. 1. Grey Relational Degree Rankings. Own study.

The ranking order of grey relational degrees obtained by grey possibility function encoding is [1–9], and the ranking order from largest to smallest obtained by the ordinal numerical encoding method is [1–9]. From the two rankings, it can be found that Evaluator 6 and Evaluator 9 rank high regardless of the method adopted, indicating their high decision quality. Similarly, Evaluator 1 and Evaluator 2 rank relatively low by both methods, indicating their low decision quality and the need for further training.

3.2 Decision Quality Assessment Model Based on Kappa Coefficient

Since the wood quality grades A, B, C, D, and NOK are ordered categories, and the severity of errors needs to be distinguished, the traditional simple Kappa coefficient consistency test is only applicable to unordered categories (such as gender and color) where the misclassification distance does not need to be considered. To better evaluate the decision quality of evaluators, this chapter adopts the weighted Kappa coefficient consistency test, which is more suitable for the situation in this paper. Its characteristics can just meet the requirements of ordered category data and the need to distinguish the severity of errors. Weighted Kappa coefficient: a statistical indicator used to evaluate the consistency between two evaluators in ordered classification tasks. Unlike the simple Kappa coefficient, it dynamically adjusts the severity of classification errors through a weight matrix, making it more suitable for scenarios where categories have a natural order. The calculation steps are as follows.

(1) **Construct the confusion matrix**

Assume there are k ordered categories in the classification task (Grade A > B > C > D > NOK), and construct a $k \times k$ confusion matrix, where n_{ij} represents the number of samples with the standard answer as category i and the evaluation result as category *j*.

(2) **Define the weight matrix W**

The weight matrix is a k × k matrix used to quantify the penalty weight for differences between categories:

$$w_{ij} = 1 - \left(\frac{|i - j|}{k - 1}\right)^2 \tag{7}$$

where $|i - j|$ represents the distance between category i and category j (e.g., the distance from A to B is 1, and from A to C is 2). A weight value closer to 1 indicates a less severe error; a value closer to 0 indicates a more severe error.

(3) **Calculate the observed consistency P_o**

The weighted observed consistency is:

$$P_o = \frac{\sum_{i=1}^{k} \sum_{j=1}^{k} w_{ij} \cdot n_{ij}}{N} \tag{8}$$

where N is the total number of samples.

(4) **Calculate the expected consistency P_e**

The weighted expected consistency based on marginal probabilities is:

$$P_e = \sum_{i=1}^{k} \sum_{j=1}^{k} w_{ij} \cdot \frac{n_{i+}}{N} \cdot \frac{n_{+j}}{N} \tag{9}$$

where n_{i+} is the total number of samples in category i in the standard answers, and n_{+j} is the total number of samples in category j in the evaluation results.

(5) **Calculate the weighted Kappa value**

$$k_w = \frac{P_o - P_e}{1 - P_e} \tag{10}$$

In the actual calculation, categorical grades such as A, AB, and B cannot be processed directly. Therefore, during data preprocessing, the matrix containing labels [A, AB, B, ...] was converted into numerical codes [1, 2, 3, ...]. For the computation, *the cohen_kappa_score function* was applied to calculate the Kappa coefficient directly. The final results are presented in Fig. 2.

According to Fig. 2, it can be seen that the evaluators show significant consistency in judging wood quality of Grade C and above, but there is obvious inconsistency in judging the boundary grades D0 and 0D (between qualified and unqualified). This is because wood has many characteristic variables, and the comprehensive evaluation of these characteristics is difficult to quantify with an accurate value. In addition, experts' subjective assumptions during decision-making lead to a greater impact of subjectivity as the number of decision characteristic variables increases. Combined with grey relational analysis, it is found that in the evaluation of quality grades such as D0 and 0D, the number of non-zero decision characteristic evaluations by the evaluators is the largest. Therefore, the evaluators cannot comprehensively consider the loss degree of each quality characteristic, and each evaluator has their own evaluation preferences and

Fig. 2. Kappa Coefficient Analysis of Each Evaluator on Different Categories. Own study.

errors. These factors together result in the situation where even with a high grey relational degree, the evaluators' classification accuracy for wood quality grades is still low.

Assuming that the acceptable average error rate for quality assessment is 20%, it can be seen from Fig. 3 that only A, BC, C, NOK, B, and D are within the acceptable threshold. Among them, the quality assessment of Grades B and D is 100% accurate, which means that in subsequent wood quality assessments, the inspection input for Grades B and D can be relatively reduced. At the same time, it can be seen that the evaluation error rates for the two quality grades DO and OD are the highest. Due to the 28 key decision characteristic variables of wood, when the wood quality grade is in the DO or OD range, the comprehensive effect of the key decision characteristic variables cannot be effectively quantified, leading to significant subjectivity. Therefore, in subsequent wood quality assessments, the review intensity for wood of grades DO and OD should be strengthened. For the ability training of quality inspectors, emphasis should be placed on training regarding the boundary between qualified and unqualified wood.

In Fig. 4, the horizontal axis represents Evaluator 1's decisions, and the vertical axis represents the standard quality assessment results. The values within the cells (e.g., 3, 1, 2) indicate the frequency of agreement between Evaluator 1's judgments and the reference (expert's) judgments for each quality grade. A higher value (represented by a darker shade) corresponds to a greater level of consistency. A value of zero (0) signifies either the absence of judgment records or complete disagreement for that particular quality grade.

For the comprehensive ability ranking of evaluators, the calculation formula adopted in this paper is: Comprehensive Score = 0.3 × Accuracy +0.25 × Kappa Coefficient +

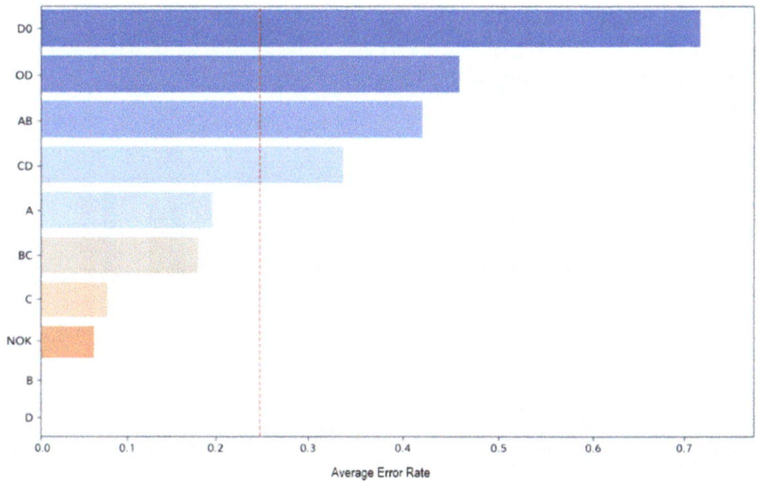

Fig. 3. Average Error Rate of Judgments for Different Quality Grades. Own study.

0.2 × NOK Recognition Accuracy +0.15 × (1 - Proportion of X Occurrence) + 0.1 × (1 - Proportion of Deviation Errors).

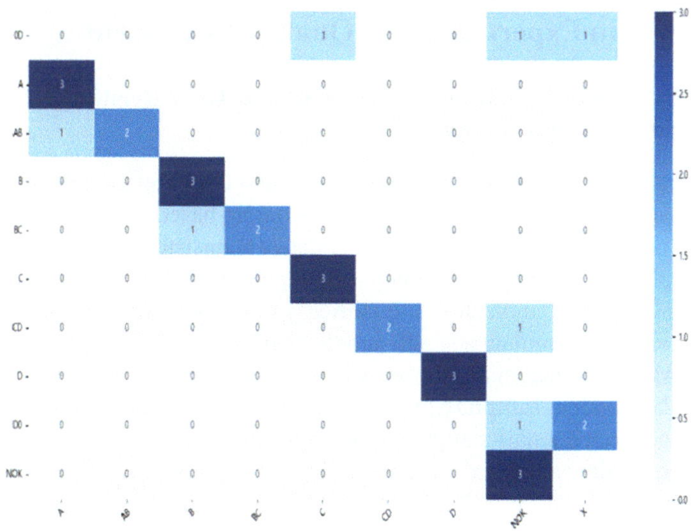

Fig. 4. Cross-Distribution between Evaluator 1's Decisions and Standard Judgments. Own study.

As shown in Fig. 5, Evaluator 6 has the highest comprehensive score (greater than 0.9), indicating the highest credibility of their wood evaluation results. The scores of the other evaluators are basically in the interval (0.8, 0.9). Evaluator 4 has the lowest comprehensive score, indicating the lowest credibility of their wood evaluation results. In future wood quality assessments, when it is necessary to combine the decision weights of each evaluator for the final decision, the weights can be assigned according to the comprehensive ability ranking of the evaluators: Evaluator 6 is assigned the highest weight, and Evaluator 4 is assigned the lowest weight.

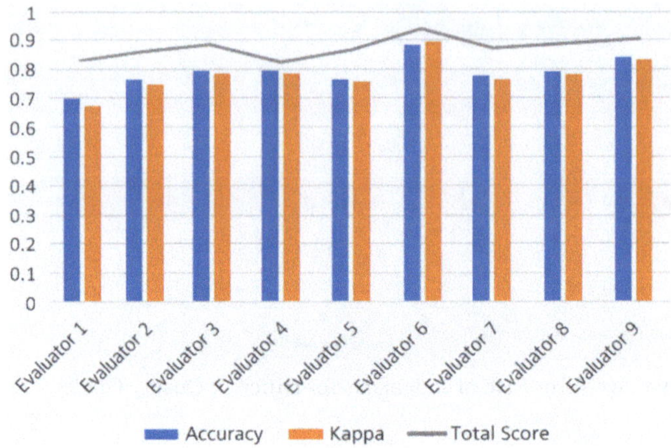

Fig. 5. Comprehensive Ability Ranking of Evaluators. Own study.

4 Comparative Analysis of Grey Relational Analysis and Kappa Coefficient in Expert Decision Quality Assessment

4.1 Reliability Analysis of Kappa Coefficient and Grey Relational Analysis Under Small Sample Sampling

To evaluate the reliability of the Kappa coefficient and Grey Relational Analysis (GRA) under small-sample conditions, a controlled experiment based on random sampling was designed and implemented. The experimental dataset consisted of inspection results for 50 wood samples assessed by nine appraisers, expressed as binary sequences (where 1 indicates 'qualified' and 0 indicates 'unqualified'), along with a corresponding reference standard sequence. To simulate a small-sample environment, subsets containing 20, 15, and 19 observations were randomly drawn from each expert's original dataset of 50 evaluations. This random sampling process was repeated 10 times to ensure robustness.

For each generated small-sample subset, both the Kappa coefficient and the grey relational degree were calculated. These small-sample results were then compared with the values obtained from the complete 50-sample dataset. To quantitatively assess the stability and consistency of the small-sample estimates relative to the full-sample benchmarks, the following statistical measures were used: average deviation (mean difference between small-sample and full-sample results), variance, standard deviation, the Kendall rank correlation coefficient (to evaluate the agreement in expert rankings between small- and full-sample results), and the coefficient of variation.

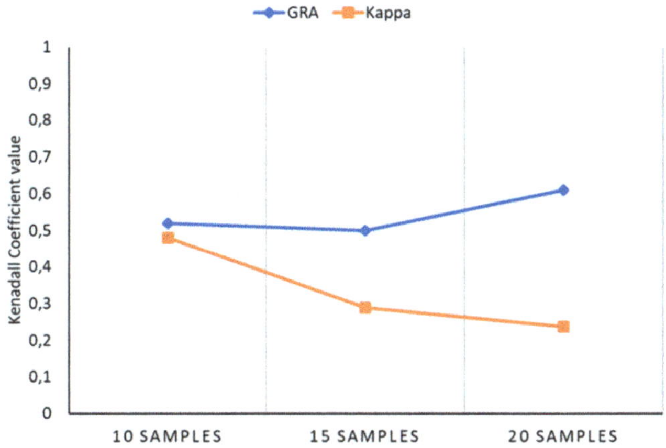

Fig. 6. Kendall Coefficients. Own study.

The experimental results demonstrate that, under small-sample conditions, the evaluation outcomes derived from grey relational analysis show a higher degree of similarity to the results obtained from the full-sample analysis. Moreover, their volatility is significantly lower than that of the Kappa coefficient consistency test. This finding indicates that when working with small-scale datasets, grey relational analysis can yield more reliable evaluation conclusions compared to the Kappa coefficient consistency test. The specific results are presented in Fig. 6 below.

4.2 Ranking of Experts' Comprehensive Decision-Making Capabilities

Overall, the assessments based on the Kappa coefficient and the grey relational degree are generally consistent. Both Evaluator 6 and Evaluator 8 rank relatively high among the nine evaluators, irrespective of whether grey relational analysis or the Kappa coefficient is applied, indicating stronger evaluation capabilities. Conversely, Evaluator 1 and Evaluator 2 consistently rank lower by both methods, reflecting weaker evaluation performance. Therefore, by integrating the results from grey relational analysis and the Kappa coefficient, it can be concluded that Evaluator 6 and Evaluator 8 exhibit higher decision quality and judgment accuracy, whereas Evaluator 1 and Evaluator 2 demonstrate lower decision quality and accuracy.

5 Conclusions

The integrated evaluation model combining grey relational analysis (GRA) and the Kappa coefficient developed in this study effectively quantifies and assesses expert decision quality across two dimensions: accuracy and stability. GRA captures the correlation between expert decisions and objective standards in terms of trend similarity and proximity, while the Kappa coefficient measures the consistency of decisions both among experts and between experts and the reference standard, thereby mitigating the influence of random factors. This approach has been successfully applied to rank the comprehensive decision-making abilities of participating experts. Furthermore, it can identify individual decision patterns and potential weaknesses in evaluating specific wood grades - such as Grade D and the D0/0D borderline grade - providing data-driven support for expert team management and tailored training.

This study is primarily based on a specific dataset provided by a Polish wood manufacturer. Consequently, the generalizability of its findings to different tree species, geographical regions, and evaluation standards requires further verification. Additionally, the limited dataset size suggests that the effectiveness of any subsequent machine-learning training would need to be validated using larger volumes of data.

Acknowledgements. This work was supported by the College of Economics and Management, Nanjing University of Aeronautics and Astronautics, Nanjing, PR China, and the Faculty of Mechanical Engineering, Poznan University of Technology, Poznan, Poland. The research presented in this article was conducted within the framework of the project "Data-driven grey system models for reliability engineering," financed by the Polish National Agency for Academic Exchange (NAWA) under grant/proposal No. BPN/BCH/2025/1/00044.

References

1. Matuszak, J., Zaleski, K., Ciecieląg, K., Skoczylas, A.: Analysis of the effectiveness of removing surface defects by brushing. Materials. **15**(21), 7833 (2022). https://doi.org/10.3390/ma15217833
2. Skoczylas, A., Zaleski, K., Matuszak, J.: Evaluation of the effectiveness of surface defect removal by slide burnishing. Appl. Sci. **15**(13), 7398 (2025). https://doi.org/10.3390/app15137398
3. Pavlenko, I., et al.: Using regression analysis for automated material selection in smart manufacturing. Mathematics. **10**(11), 1888 (2022)
4. Jozsa, L.A., Middleton, G.R.: A discussion of wood quality attributes and their practical implications. Forintek Canada Corp. (1994)
5. Kozierkiewicz-Hetmańska, A.: The analysis of expert opinions' consensus quality. Inf. Fusion. **34**, 80–86 (2017). https://doi.org/10.1016/j.inffus.2016.06.005
6. Dror, I.E., Charlton, D.: Why experts make errors. J. Forensic Identif. **56** (2006)
7. Geva, T., Saar-Tsechansky, M.: Who is a better decision maker? Data-driven expert ranking under unobserved quality. Prod. Oper. Manag. **30**, 127–144 (2021). https://doi.org/10.1111/poms.13260
8. Liu, S.: Grey Systems Analysis. Springer Nature Singapore, Singapore (2025). https://doi.org/10.1007/978-981-97-8727-2
9. Wu, H., Li, Y., Han, X., Hu, A.: Model validation and visualization techniques of Grey relational analysis. J. Grey Syst. **37**, 100–112 (2025)
10. Hakan, H., Canlıer, E.C., Özarı, Ç., Can, E.N.: Modeling and predicting the socio-economic performance of countries using grey relational analysis and K-NN algorithm. J. Grey Syst. **36**, 40–50 (2024)
11. Zhou, Z., et al.: Advanced CRITIC–GRA–GMM model with multiple restart simulation for assuaging decision uncertainty: an application to transport safety engineering for OECD members. Adv. Eng. Informatics. **60**, 102373 (2024). https://doi.org/10.1016/j.aei.2024.102373
12. Zhang, N., Zhou, J., Mao, Z., Chen, D.: Quantum interference-grey correlation (QGRA) modelling of UAV combat effectiveness assessment. J. Grey Syst. **37** (2025)

13. Diering, M., Dyczkowski, K., Hamrol, A.: Rating system analysis in quality engineering based on fuzzy similarity. J. Mult. Valued Logic Soft Comput. **32**(3–4), 277–292 (2019)
14. Kraemer, H.C.: Extension of the kappa coefficient. Biometrics. **36**, 207 (1980). https://doi.org/10.2307/2529972
15. Aziz, G., Minallah, N., Saeed, A., Frnda, J., Khan, W.: Remote sensing based forest cover classification using machine learning. Sci. Rep. **14**, 69 (2024). https://doi.org/10.1038/s41598-023-50863-1
16. Wan, T.A.N.G., Jun, H.U., Pan, W.U., Hua, H.E.: Kappa coefficient: a popular measure of rater agreement. Shanghai Arch. Psychiatry 27, (2015). 10.11919/j.issn.1002-0829.215010.
17. Kujawińska, A., Diering, M., Rogalewicz, M., Żywicki, K., Hetman, Ł.: Soft modelling-based methodology of raw material waste estimation. In: Burduk, A., Mazurkiewicz, D. (eds.) Intelligent Systems in Production Engineering and Maintenance. Advances in Intelligent Systems and Computing, vol. 637. Springer, Cham (2018). https://doi.org/10.1007/978-3-319-64465-3_39
18. Kujawińska, A., Diering, M., et al.: Methodology supporting the planning of machining allowances in the wood industry. In: Pérez García, H., et al. (eds.) Advances in Intelligent Systems and Computing, vol. 649. Springer, Cham (2017). https://doi.org/10.1007/978-3-319-67180-2_33

Integration of Digital Inspection Cards with QMS/MES in the Automotive Parts Remanufacturing Process–A Case Study

Łukasz Wójcik, Arkadiusz Gola(✉), and Jakub Pizoń

Faculty of Mechanical Engineering, Lublin University of Technology, Lublin, Poland
a.gola@pollub.pl

Abstract. The article addresses the issue of quality assurance in remanufacturing, embedded in the circular economy paradigm. In particular, it presents the use of digital inspection cards integrated with MES/QMS, which standardize OK/NOK/Acceptable decisions, ensure full traceability, and feed quality analytics in a company involved in the regeneration of parts for the automotive industry. The case study presented covers the regeneration of brake calipers with two critical nodes: sorting (rejection rate of 31.73%; dominated by body corrosion, exceeded working length, and mechanical damage) and final testing (actual FAIL 1.88% of production; in the event log, 35.37% of actual rejects after excluding the "OK/for re-regeneration" items; >50% scrap due to corrosion of the body/handles). The implementation of digital inspection cards with data records by operators and quality engineers into MES/QMS improved the consistency of assessments and reduced inspection time. The results confirm that strengthening quality in regeneration accelerates the achievement of circular economy goals, combining environmental benefits with reduced quality costs and greater process predictability.

Keywords: Remanufacturing · Quality · Digital Inspection Cards · Quality Management System · Manufacturing Execution System

1 Introduction

The declining availability of natural resources and growing environmental pressure are prompting industry to move away from the linear "take-make-use-dispose" model in favor of a circular economy [1]. In this approach, remanufacturing is a key mechanism for extending the life cycle of products, reducing the consumption of primary raw materials and reducing waste. The entire process aims to restore the product to a condition equivalent to new in terms of functionality and reliability [2]. The economic and environmental appeal of this approach is particularly evident for complex mechanical and electromechanical products, where the value of recovered cores is significant [3].

However, the specific nature of the input stream in remanufacturing – the variability in quality and uncertainty regarding the condition of end-of-life (EoL) products – generates unique quality challenges [4]. The cores come from operations of varying intensity and

M. Hryb et al. (Eds.): MANUFACTURING 2026, LNME, pp. 220–232, 2026.
https://doi.org/10.1007/978-3-032-23038-6_16

conditions, which makes it difficult to reliably assess them before starting the process and complicates the planning of the time and cost of restoring functionality. In practice, this means greater sensitivity to classification errors (false acceptances and rejections), the need for precise management of "fit/repair/reject" decisions, and the need for strict traceability at the batch and item level [5].

At the same time, the digitisation of processes is accelerating within the framework of Industry 4.0 and Industry 5.0. QMS (Quality Management Systems), MES (Manufacturing Execution Systems), the Internet of Things, data analytics, vision systems and machine learning now form an ecosystem that can significantly improve the consistency and effectiveness of quality assessment in regeneration [6]. At the heart of this ecosystem are digital inspection cards – structured inspection instructions that standardise assessment criteria, enforce completeness of entries (defect codes, location, decision), enable the attachment of photographic documentation and immediately feed into the quality data repository. The integration of cards with QMS/MES ensures the ongoing flow of information between technology hubs and quality functions, supports layered audits, and provides data for SPC (Statistical Process Control) and Pareto analyses [7, 8].

Quality in regeneration, understood in this way, is not limited to detecting non-compliance, but includes designing the information needed to make the right decision and teaching the organisation on the basis of data [9]. Digital inspection cards – metrologically validated (MSA – Measurement Systems Analysis) and embedded in the versioning process – become a carrier of expert knowledge, a tool for reducing variability in assessments between inspectors, and a source of reliable indicators [10]. As a result, organisations can more effectively balance environmental, economic and quality objectives, and remanufacturing itself – as a pillar of the circular economy – gains in predictability and scalability [11].

The aim of this article is to present the use of digital inspection cards in remanufacturing processes, with particular emphasis on their integration with MES/QMS systems and their impact on the consistency of attribute decisions. The article illustrates how to design and implement cards (criteria, defect dictionaries, photographic examples) and shows, using the example of brake caliper regeneration, how digital cards contribute to improving the quality and predictability of processes within the circular economy.

2 The Role of Remanufacturing in the Circular Economy

The circular economy is a model in which the value of products, components and materials is maintained in circulation for as long as possible through design for durability and repairability, reuse, refurbishment, remanufacturing and recycling. Its aim is to minimise waste generation and the consumption of primary resources, as well as to reduce emissions and environmental pressure [12]. Consequently, the circular economy is a modern economic model that assumes systemic regeneration and maximum use of resources by designing products with reuse, remanufacturing and recycling in mind. Unlike the traditional linear model ('take, make, dispose'), the circular economy focuses on minimising material consumption, eliminating waste and using renewable energy [13, 14]. Key principles include maximising the number of product life cycles, using, materials in a cascading manner (e.g. cotton passing through different stages of use) and maintaining high material quality by keeping them in 'clean cycles'. The circular economy contributes to reducing material costs, mitigating the risk associated with raw material prices, and can lead to significant savings, reaching hundreds of billions of dollars annually [15]. This model promotes innovation, improves competitiveness and responds to the challenges of limited resources and market volatility [16, 17]. The circular economy system diagram is presented in Fig. 1.

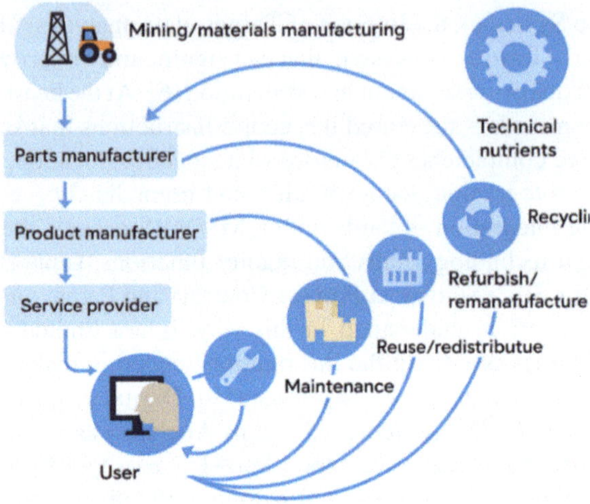

Fig. 1. Circular economy system diagram (own work based on [11]).

The key principles of the circular economy are [18]:

- value retention: priority for actions with the highest value retention (keeping the product/component 'as close as possible' to its usable condition),
- closed-loop design: standardisation, modularity, easy disassembly, availability of parts, functional restoration paths,
- life cycle management: data-driven decisions (LCA, carbon footprint, material footprint) and re-market paths,
- supply chain integration: OEM–remanufacturing–service–recycling cooperation, traceability and quality tracking.

Remanufacturing plays a key role in the circular economy due to its significant contribution to recovering value from end-of-life products. The remanufacturing process, which involves replacing or reprocessing used components in order to restore them to full functionality, significantly reduces the consumption of natural resources and the generation of waste in the production cycle. As a result, regeneration not only improves the environmental performance of companies, but also strengthens their market position [19, 20].

As the literature indicates, remanufacturing as an economic sector brings numerous social, environmental and economic benefits. It promotes sustainable development that takes into account three key aspects: people, the environment and company profits [21, 22]. Despite the obvious environmental benefits that meet the requirements of extended producer responsibility regulations, regeneration also offers a number of other significant advantages. Research shows that by recovering used products, manufacturers can obtain key feedback on the reliability and durability of their products, thus diagnosing the most common defects, faults or problems affecting parts and components, which provides valuable support in improving production processes [23, 24].

For consumers, this means access to high-quality end products at competitive prices, while for manufacturers, regeneration is a tool for protecting intellectual property and brand image. In addition, with the industry growing rapidly, their availability is easier. Sometimes there are situations where it is only possible to purchase remanufactured parts because their new counterparts have been withdrawn from the market or their suppliers no longer exist.

From an economic point of view, remanufacturing enables products to be placed on the market at prices 40–80% lower than new products, which contributes to increased availability and profitability of the offer [25]. However, it is important that regeneration must be economically viable for businesses [26]. This is why it is particularly useful in the automotive industry. For example, product regeneration consumes only 15–20% of the total energy and materials needed to manufacture a new component. This is because original parts can be reused in the remanufacturing proces [27]. Research shows that regeneration can further reduce air pollutant emissions by 80% [28]. For these reasons, the regenerative market is constantly growing. It is estimated that by 2030, its value in Europe could reach €100 billion [29].

3 The Course and Role of Quality in the Regeneration Process

The remanufacturing process begins with the acceptance of the product and its unambiguous identification. At this stage, the catalogue and serial numbers are verified, the component is checked for completeness, and the declared damage symptoms are verified. After preliminary qualification, the unit is disassembled into its component parts using dedicated tools and instructions to avoid secondary damage. The parts are sorted into groups: those that can be restored, those that must be replaced, and those that are sent for recycling. Replaceable components usually include seals, bearings, rubber and friction elements – these components cannot be reused [30–32]. Process fluids, such as oils and working fluids, are separated and disposed of in accordance with environmental requirements, and the completeness of parts packages is confirmed in the dismantling documentation.

Cleaning is a crucial step, as it determines the reliability of subsequent defect detection. The choice of method – from water-alkaline cleaning and ultrasonic cleaning to abrasive blasting – depends on the material and geometry of the part. The aim of this stage is to remove contamination, corrosion and wear products without damaging the surface layer and without introducing measurement artefacts. For precision systems, technical cleanliness requirements are maintained, and the effectiveness of cleaning is confirmed by testing for contamination residues [33].

The cleaned parts undergo comprehensive inspection and sorting. This includes visual assessment according to acceptance criteria, non-destructive testing (penetration, magnetic, ultrasonic or eddy current), geometric and roughness metrology, and, if necessary, material analysis. Repair operations restore the dimensions, geometry or properties of the surface layer to the level required by the specification. Classic subtractive machining (turning, milling, grinding, honing), permanent joining (welding, brazing) and thermal and thermochemical treatments are used to restore the hardness and load-bearing capacity of the layer [34].

The next stage is final assembly, where all the reconditioned components are assembled into a complete unit, which is then sent to the End-of-Line test station, where it undergoes functional verification – from leak and flow characteristic tests, through torque, clearances, noise and vibration, to electrical tests for mechatronic components. Only when documented compliance with the criteria is achieved is the product released [35]. The main processes used during regeneration and described above are presented in Fig. 2.

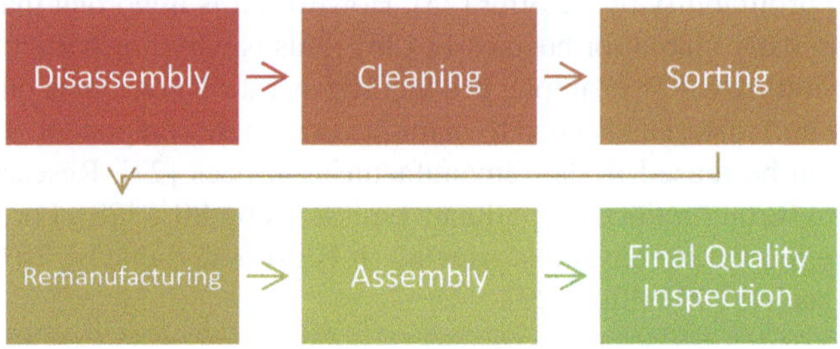

Fig. 2. Simplified remanufacturing diagram (own work based on: [35]).

Quality in regeneration differs significantly from quality in primary production due to the variability of the input condition and uncertainty regarding the further machinability of components [36]. The source of risk lies in attribute decisions made on the basis of visual inspection and functional tests: false acceptances result in defects 'escaping' to the customer, while false rejections generate unnecessary costs and prolong the remanufacturing cycle [37]. Therefore, the quality system should combine precise evaluation criteria, validated inspector competencies, and consistent data flow supported by QMS/MES [38, 39].

The basis for standardisation is a visual inspection manual, which defines: the scope of assessment, the sequence of steps, defect classes (critical/major/minor), measurable acceptance criteria (wear limit, acceptable surface area of defect, tolerances), a dictionary of defect codes and photographic references ("good/bad"). Such instructions should enforce the completeness of the record (item/batch identifier, operator, decision, defect code, location, photo). The inspection sequence should be logically divided into the following stages: entry (core qualification), special processes (e.g. machining, cladding), FQC (Final Quality Control) [40].

In order to reduce the variability of assessments, it is necessary to perform a Measurement System Analysis (MSA) for attribute data. The MSA results are linked to the training and certification plan for inspectors. At the same time, the inspection station should be organised in an appropriate manner in terms of lighting, background, magnification and ergonomics, as these conditions significantly affect the detection of defects.

The effectiveness of the quality system is measured by a set of KPIs (Key Performance Indicators) such as:

- FPY (First Pass Yield),
- DPMO (Defects Per Million Opportunities)/PPM (Parts Per Million),
- inspection time/unit,
- percentage of rework/scrap,
- after-sales complaint rates

Data should flow automatically from operator reports to QMS/MES, enabling trend monitoring, Pareto analysis for defect structure, and rapid action initiation. At the preventive level, PFMEA (Process Failure Mode and Effects Analysis) should identify high-risk points, and the Control Plan should map them to specific control operations and visual control instruction requirements. Additional safeguards include audit "last gates" and LPA (Layered Process Audit) audits, which verify the actual application of instructions at the workstation.

A quality system designed in this way – combining clear instructions, proven MSA methods and continuous data flow – increases the repeatability of decisions, shortens inspection times and reduces quality costs, creating a solid foundation for scaling up regeneration as a pillar of the circular economy.

4 Case Study: Digital Inspection Cards and Integration with QMS/MES in Brake Caliper Regeneration

This chapter presents a case study of the implementation of digital inspection cards integrated with QMS/MES systems in the brake caliper regeneration process. An example of a caliper before and after regeneration is shown in Fig. 3. The analysis focuses on two key slots that have the greatest impact on the quality and costs of the process:

1. Sorting (core qualification) – the first quality 'gate', where decisions are made about further processing of the component (OK/repair/rejection).
2. End-of-Line test – the final 'gate' that verifies compliance with functional and safety requirements before the product is released

The aim of the study is to show how structured digital inspection cards – with a set of fields (item/batch identification, defect codes, location, photo documentation), entry validation rules and versioning – and full data integration with QMS/MES affect:

- information flow,
- key process indicators, such as scrap percentage and defect structure (Pareto).

The source data comes from a company involved in the remanufacturing of automotive components. The following subsections document: the process flow, the structure of digital inspection cards, the data flow map (visual inspection instructions ↔ QMS/MES), results and operational conclusions from a quality perspective.

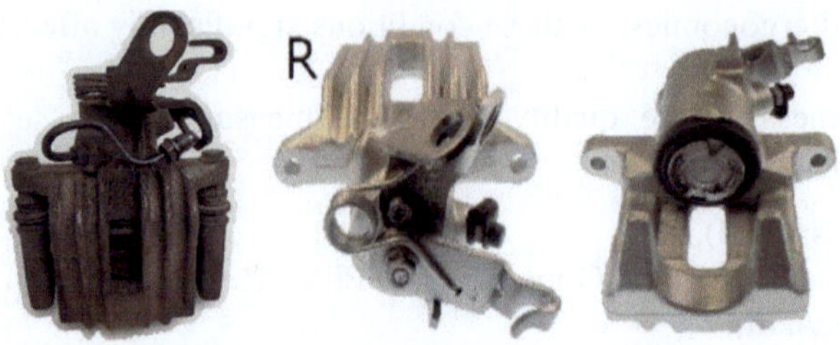

Fig. 3. Brake caliper before and after regeneration.

4.1 The Brake Caliper Regeneration Process

The brake caliper remanufacturing process begins with dismantling the assembly into its component parts. The parts are then washed and cleaned. After cleaning, they are sorted and verified for further usability and compliance with requirements. Qualified parts are reconditioned by turning, milling, inserting bushings, etc. Unevenness and minor corrosion defects are levelled out at the filling station. Before assembly, the parts are washed again to remove any remaining material and contaminants.

The next step is pre-assembly, i.e. preliminary assembly of selected components and completion of the parts required for final assembly. The main assembly of the clamp takes place during the assembly stage, after which each unit undergoes a final test (End of Line) to verify its functionality and tightness before release to the customer. The course of this process is presented in Fig. 4.

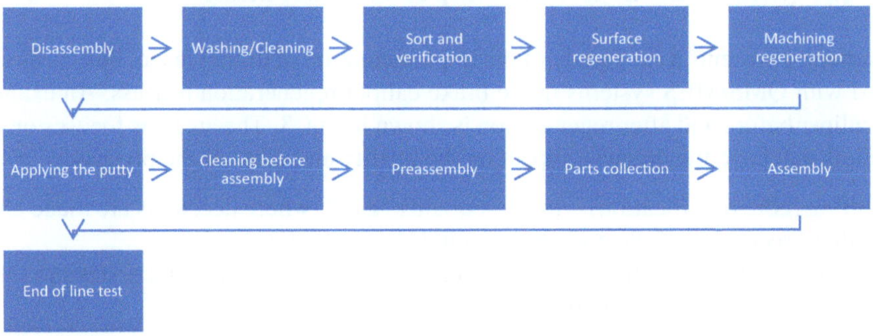

Fig. 4. The brake caliper regeneration process.

4.2 Digital Inspection Cards – Content, Standard and Use

The digital inspection card contains all the information needed to make an unambiguous "OK/NOK/Acceptable" assessment for a given item. The header contains the reference and order identification (item/batch ID, station, date-time, operator) and the card version. The core consists of control points with described criteria (tolerances, thresholds, conditional acceptance conditions), a dictionary of defect codes (name, description, location/side, default decision) and photo references: examples of OK, Acceptable (with conditions) and NOK Figs. 5 and 6.

The operator in the registration section reports in the MES system: for each feature (OK/NOK/Acceptable), a defect code is assigned, comments are added, and the inspection time is recorded automatically. The data is simultaneously sent to the QMS, where it is analysed; in the event of deviations (e.g. an increase in the % of scrap), the system sends notifications to quality engineers.

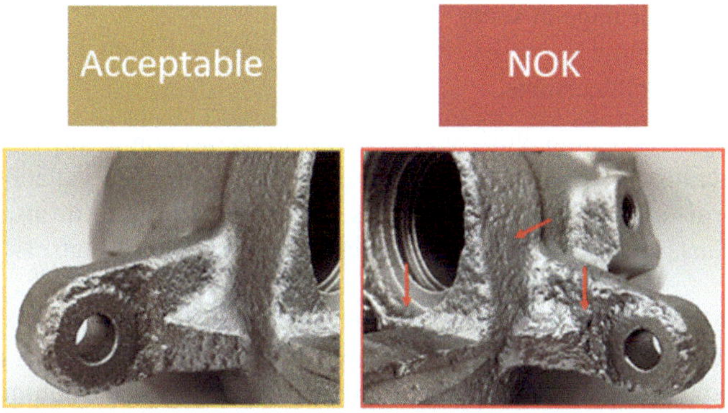

Fig. 5. Examples of defects - corrosion of the mounting ears (source: own study).

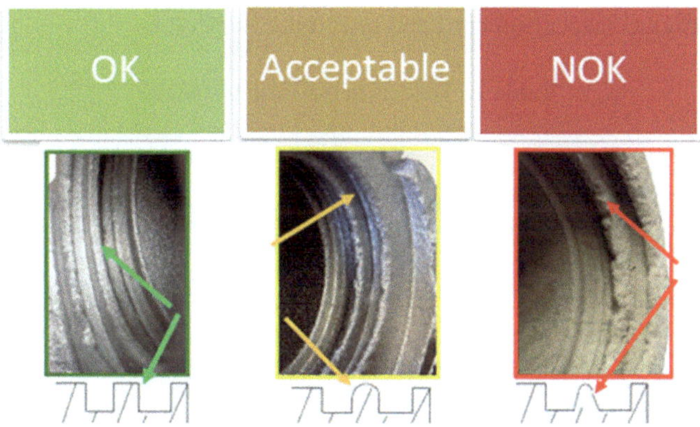

Fig. 6. Examples of defects - corrosion of the dust-boot seat

4.3 Results and Their Interpretation

Data from the brake caliper regeneration process is recorded in two ways: by production employees performing operations at individual stations and by quality controllers who conduct secondary verifications and compliance audits. Each quality event (OK/repair/reject decision, defect code, defect location, comments, possible photographic documentation) is recorded in the MES system, which is responsible for classifying entries, assigning them a process context (station, operation, order number, shift, operator, time) and linking them to the current version of the digital inspection card. This study analyses data from two key points in the process: sorting (core/component qualification) and final testing (End of Line), which are the first and last quality gates, respectively.

The recorded information enables piece-by-piece traceability and multidimensional analysis. In particular, using the sorting and EOL database, it is possible to:

- identify types of defects, count their number and percentage share in the population (Pareto structure),
- calculate the percentage of scrap
- track trends over time and assess the impact of changes in digital inspection cards on the repeatability of decisions.

The purpose of collecting this data is not only to record non-conformities, but above all to make operational decisions and improve the process: rapid escalation of incidents to QMS, prioritisation of corrective actions, updating of criteria in inspection sheets and control plans, targeted training of inspectors and production staff, as well as better planning.

During the analysed period, 6,325 brake pads were assessed at the sorting station. 2,007 items were classified as NOK, which gives a scrap rate of 31.73% at the start of the process. All defects listed in Table 1 are irreparable at this stage and automatically result in a "reject" decision. The Pareto structure shows that body defects dominate: severe corrosion, exceeded working length and mechanical damage. These three categories account for nearly 58% of all rejections and should be the main focus of refining the criteria in the digital inspection card and retraining inspectors.

Table 1. Defects at the sorting station

No.	Defect name	Quantity [pcs]	Participation [%]
1	Severe corrosion of the housing	647	32,13
2	Exceeded working length of the housing	297	14,75
3	Mechanical damage to the housing	216	10,72
4	Corrosion of the centers	182	9,04
5	Corrosion of the dust-boot seat	166	8,24
6	Corrosion of the mounting ears	110	5,46
7	Cylinder/bore corrosion	75	3,72
8	Improper removal of bearings and bushings	75	3,72
9	Oversize center holes	74	3,67
10	Deformed mounting ear	67	3,33
11	Non-original housing	27	1,34
12	Corrosion of the flange/groove for the sealing ring	26	1,29
13	Corrosion of the housing guides	19	0,94
14	Peened/corroded surface around the pressure port/vent	15	0,74
15	Cracked housing	6	0,3
16	Deformed cylinder/bore	5	0,25

A total of 4,318 units were tested in the end-of-life (EoL) test. A separate EoL event log (229 cases) – Table 2 – shows that items marked as No. 1, 2, 5, 6, 7, 9, 12, 16 are not actual rejects – these are "OK" cases or cases eligible for re-regeneration (e.g. "Part OK – credit note..."). After excluding them, there are 81 actual rejects. Interpretatively, this gives two complementary indicators: 35.37% of rejects among the events in the EoL table alone (81/229) and 1.88% of rejects relative to the entire population tested on EoL (81/4,318).

Table 2. Defects at the final test station

No.	Defect name	Quantity [pcs]	Participation [%]
1	Part OK – credit note for a "non-Bosch" customer	52	22,71
2	Incorrect calculation of the batch quantity	35	15,28
3	Severe corrosion of the housing	26	11,35
4	Corrosion of the mounting ears	18	7,86
5	Leak at the pressure supply port	17	7,42
6	Incorrect piece count reported from the previous station	16	6,99
7	Work order Card "0": damaged supply-port thread (insert)	11	4,8
8	Over-ground shaft centers	9	3,93
9	Work order Card "0": leak at/around the sector-shaft	9	3,93
10	Improperly installed thread insert	8	3,49
11	Damaged supply-port thread (insert cannot be installed)	7	3,06
12	Work order Card "0": leak from the vent/breather	7	3,06
13	Mechanical damage to the housing	5	2,18
14	Assembly not possible	4	1,75
15	Test result: FAIL	4	1,75
16	Damaged supply-port thread	1	0,44

5 Conclusions

The implementation of digital inspection cards integrated with MES/QMS systems enables consistency in attribute decisions and provides full traceability at the individual item level. A case study of brake caliper regeneration indicates a high concentration of non-conformities: sorting is dominated by body defects (corrosion, exceeded working length, mechanical damage), which together account for approximately 57.6% of rejects. The total rejection rate during sorting was 31.73% (2007/6325). At the EoL station, the actual share of rejections in total production was low (1.88%; 81/4318), but among the events recorded in the EoL log (n = 229), 35.37% were actual non-conformities after excluding the 'OK/for re-regeneration' category. More than half of the EoL rejects were generated by severe corrosion of the body and corrosion of the mounting brackets. These results confirm the validity of focusing improvement efforts on a narrow group of causes with the greatest impact (Pareto principle) and maintaining dual EoL reporting (incidence among events vs. incidence relative to total production), which reduces the risk of misinterpretation.

In practical terms, it is advisable to refine digital inspection cards in terms of the five most common causes of rejection, with clear acceptance criteria, standardised "OK/Acceptable/NOK" photo references and a requirement for photographic documentation in borderline cases. In addition, it is suggested that early checking for corrosion of the body be introduced at the entrance to the process (pre-sort). The third aspect that was highlighted is the appropriateness of using simple solutions such as poka-yoke for threads and inserts (short "go/no-go" tests before assembly), as well as a formal "handover check" between sockets. It is also important to establish alarm thresholds in the QMS for % scrap and the share of key codes, with automatic notification of corrective action owners and continuous Pareto Top-10 reporting.

The identified priorities allow us to formulate operational objectives: reduction of the total share of the three main categories in sorting by $\geq$20% within 3 months (decrease in % scrap from 31.73% to $\leq$25%), maintaining the % FAIL at EoL $\leq$ 2.0% for the entire production and reducing the share of the two main causes of EoL (corrosion of the body/handles) by $\geq$30% within 90 days.

The limitations of the study should be emphasised. Firstly, the variability in core quality (EoL) depends on the history of use and recovery logistics, which makes direct period-to-period comparisons difficult. Secondly, some of the categories recorded at EoL are procedural in nature (e.g. discrepancies in batch sizes), which requires them to be consistently excluded from production scrap measures and treated separately in root cause analysis. Thirdly, the impact of seasonality (storage conditions, humidity) and organisational changes (staff turnover, training) may have affected the observed indicators regardless of the records made using digital inspection cards. For these reasons, the results, although operationally convincing, should be interpreted taking into account the context of the process and the quality of the data.

Future research plans include analyses extended to predictive modelling of rework load (using process characteristics, core sources and inspection card versions), as well as a quantitative assessment of the environmental impact of scrap reduction. In light of the results obtained, the integration of the digital data quality inspection card (MES/QMS) work standard and the short decision loop is the most effective path to sustainable quality improvement in remanufacturing processes.

References

1. European Investment Bank: The EIB Circular Economy Guide. Supporting the Circular Transition, European Investment Bank (2019)
2. Piechowski, M., Meller, A.: Use of KPI indicators to optimize production process performance – case study. Distribut. Comput. Artif. Intell. **3**, 3–12 (2023)
3. Schau, E.M., Traverso, M., Finkbeiner, M.: Life cycle approach to sustainability assessment: a case study of remanufactured alternators. J. Remanuf. **2**(1), 1–14 (2012). https://doi.org/10.1186/2210-4690-2-5

4. Ponte, B., Cannella, S., Dominguez, R., Naim, M.M., Syntetos, A.A.: Quality grading of returns and the dynamics of remanufacturing. Int. J. Prod. Econ. **236**, 108129 (2021). https://doi.org/10.1016/j.ijpe.2021.108129
5. El Saadany, A.M.A., Jaber, M.Y.: A production/remanufacturing inventory model with price and quality dependent return rate. Comput. Ind. Eng. **58**(3), 352–362 (2010). https://doi.org/10.1016/j.cie.2009.01.017
6. Jamwal, A., Agrawal, R., Sharma, M.: Deep learning for manufacturing sustainability: models, applications in industry 4.0 and implications. Int. J. Inf. Manag. Data Insights. **2**(2), 100107 (2022). https://doi.org/10.1016/j.jjimei.2022.100107
7. Azamfirei, V., Psarommatis, F., Lagrosen, Y.: Application of automation for in-line quality inspection, a zero-defect manufacturing approach. J. Manuf. Syst. **67**, 1–22 (2023). https://doi.org/10.1016/j.jmsy.2022.12.010
8. Meller, A., Piechowski, M., Gola, A.: Indicators of hierarchical structure model supporting the production management process – a framework. Intell. Syst. Product. Eng. Maint. **356**, 356–367 (2023)
9. Ślusarczyk, B.: Industry 4.0 – are we ready? Polish. J. Manag. Stud. **17**(1), 232–248 (2018). https://doi.org/10.17512/pjms.2018.17.1.19
10. Subramoniam, R., Sundin, E., Subramoniam, S.: Sustainability in remanufacturing. Sustainability. **13**, 1–23 (2021)
11. Lieder, M., Rashid, A.: Towards circular economy implementation: a comprehensive review in context of manufacturing industry. J. Clean. Prod. **115**, 36–51 (2016). https://doi.org/10.1016/J.JCLEPRO.2015.12.042
12. Linder, M., Williander, M.: Circular business model innovation: inherent uncertainties. Bus. Strateg. Environ. **26**(2), 182–196 (2017). https://doi.org/10.1002/bse.1906
13. Tsolakis, N., Harrington, T.S., Srai, J.S.: Digital supply network design: a circular economy 4.0 decision-making system for real-world challenges. Prod. Plan. Control. **34**(10), 941–966 (2023). https://doi.org/10.1080/09537287.2021.1980907
14. Llerena, D.: Green consumer behaviour: an experimental analysis. Bus. Strateg. Environ. **20**(6), 408–420 (2011)
15. Saboori, A., Aversa, A., Marchese, G., Biamino, S., Lombardi, M., Fino, P.: Application of directed energy deposition-based additive manufacturing in repair. Appl. Sci. **9**(16), 3316 (2019). https://doi.org/10.3390/app9163316
16. EMAF: Towards the circular economy. J. Ind. Ecol. **2**(1), 23–44 (2013)
17. Camacho-Otero, J., Boks, C., Pettersen, I.N.: Consumption in the circular economy: a literature review. Sustainability. **10**(8), 2758 (2018). https://doi.org/10.3390/su10082758
18. Prieto-Sandoval, V., Jaca, C., Ormazabal, M.: Towards a consensus on the circular economy. J. Clean. Prod. **179**, 605–615 (2018). https://doi.org/10.1016/j.jclepro.2017.12.224
19. Miao, Z., Mao, H., Fu, K., Wang, Y.: Remanufacturing with trade-ins under carbon regulations. Comput. Oper. Res. **89**, 253–268 (2018). https://doi.org/10.1016/j.cor.2016.03.014
20. Seitz, M.A.: A critical assessment of motives for product recovery: the case of engine remanufacturing. J. Clean. Prod. **15**(11–12), 1147–1157 (2007). https://doi.org/10.1016/j.jclepro.2006.05.029
21. Pizoń, J., Gola, A.: The meaning and directions of development of personalized production in the era of industry 4.0 and industry 5.0. Lecture notes. Mech. Eng. **1**, 1–12 (2023). https://doi.org/10.1007/978-3-031-09360-9_1
22. Nußholz, J.L.K.: Circular business models: defining a concept and framing an emerging research field. Sustainability. **9**(10), 1810 (2017). https://doi.org/10.3390/su9101810
23. Gunasekara, H., Gamage, J., Punchihewa, H.: Remanufacture for sustainability: barriers and solutions to promote automotive remanufacturing. Procedia Manuf. **43**, 606–613 (2020). https://doi.org/10.1016/j.promfg.2020.02.146

24. Morrow, W.R., Qi, H., Kim, I., Mazumder, J., Skerlos, S.J.: Environmental aspects of laser-based and conventional tool and die manufacturing. J. Clean. Prod. **15**(10), 932–943 (2007). https://doi.org/10.1016/j.jclepro.2005.11.030
25. Wei, S., Tang, O., Sundin, E.: Core (product) acquisition management for remanufacturing: a review. J. Remanuf. **5**(1), 1–14 (2015). https://doi.org/10.1186/s13243-015-0014-7
26. Relich, M., Nielsen, I., Gola, A.: Reducing the total product cost at the product design stage. Appl. Sci. **12**(4), 1921 (2022). https://doi.org/10.3390/app12041921
27. Zhou, Q., Meng, C., Sheu, J.B., Yuen, K.F.: Remanufacturing mode and strategic decision: a game-theoretic approach. Int. J. Prod. Econ. **260**, 108841 (2023). https://doi.org/10.1016/j.ijpe.2023.108841
28. Si, X., Wang, C., Mao, H., Liu, C.: Remanufacturing processing path decision-making method considering low-carbon and quality. Int. J. Adv. Manuf. Technol. **1**, 6131–6146 (2025). https://doi.org/10.1007/s00170-025-17017-1
29. Caterino, M., Fera, M., Macchiaroli, R., Pham, D.T.: Cloud remanufacturing: remanufacturing enhanced through cloud technologies. J. Manuf. Syst. **64**, 133–148 (2022). https://doi.org/10.1016/j.jmsy.2022.06.003
30. Andrew-Munot, M., Ibrahim, R.N.: Remanufacturing process and its challenges. J. Mech. Eng. Sci. **4**, 488–495 (2013). https://doi.org/10.15282/jmes.4.2013.13.0046
31. Pizoń, J., Wójcik, Ł., Gola, A., Kański, Ł., Nielsen, I.: Autonomous mobile robots in automotive remanufacturing: a case study for intra-logistics support. Adv. Sci. Technol. Res. J. **18**(1), 213–230 (2024). https://doi.org/10.12913/22998624/177398
32. Wlazlak, P., Johansen, K.: Development of remanufacturing systems at original equipment remanufacturers. Int. J. Interact. Design Manuf. **19**(12), 8641–8661 (2025). https://doi.org/10.1007/s12008-025-02314-6
33. Hu, S.J., et al.: Assembly system design and operations for product variety. CIRP Ann. Manuf. Technol. **60**(2), 715–733 (2011). https://doi.org/10.1016/j.cirp.2011.05.004
34. Jiang, Z., Fan, Z., Sutherland, J.W., Zhang, H., Zhang, X.: Development of an optimal method for remanufacturing process plan selection. Int. J. Adv. Manuf. Technol. **72**(9–12), 1551–1558 (2014). https://doi.org/10.1007/s00170-014-5783-x
35. Litwin, P., Gola, A., Wójcik, Ł., Cioch, M.: Optimization of the flow of parts in the process of brake caliper regeneration using the system dynamics method. PRO. **12**(1), 16 (2024). https://doi.org/10.3390/pr12010016
36. Wu, Z., Liang, C., Shen, Z., Zhou, W.: Quality control in remanufacturing process: theoretical framework, control strategy and application review. Int. J. Adv. Manuf. Technol. **140**(1–2), 517–543 (2025). https://doi.org/10.1007/s00170-025-16343-8
37. Aksoy, H.K., Gupta, S.M.: Buffer allocation plan for a remanufacturing cell. Comput. Ind. Eng. **48**(3), 657–677 (2005). https://doi.org/10.1016/j.cie.2003.03.007
38. Wójcik, Ł., Gola, A.: Working time standardization in the assembly process of regenerated elements using MES system and timing method. Lecture Notes Networks Syst. **5**, 1–10 (2023). https://doi.org/10.1007/978-3-031-38318-2_5
39. Ibrahim, R.: Digital quality management systems: benefits and challenges. Proc. Eng. Sci. **1**(2), 163–172 (2019). https://doi.org/10.24874/PES01.02.015
40. Goodall, P., Sharpe, R., West, A.: A data-driven simulation to support remanufacturing operations. Comput. Ind. **105**, 48–60 (2019). https://doi.org/10.1016/j.compind.2018.11.001

Automation of the Cylindrical Hole Measurement Through the Application of a Multi-criteria Tool Selection Matrix and the Extraction of Geometric Data from the CAD Model

Danuta Owczarek[1(✉)], Ksenia Ostrowska[1], Ilona Jędrzejak[1], Csaba Felhő[2], and Michał Stoliński[1]

[1] Cracow University of Technology, Kraków 31-155, Poland
danuta.owczarek@pk.edu.pl

[2] Institute of Manufacturing Science, University of Miskolc, H-3515 Miskolc-Egyetemváros, Hungary

Abstract. The article addresses the common task of precise hole measurement in machine and equipment manufacturing, the automotive sector, aviation, and the defense industry. Attention is drawn to the challenges associated with this process, particularly when the hole depth to diameter ratio is significant, which considerably complicates the acquisition of accurate measurement data. Previously employed measurement solutions are also presented, including the use of coordinate measurement techniques and selected optical systems, such as triangulation-based laser systems, structured light systems, and computed tomography. Example results are shown, and the observed regularities during hole measurements using these systems are discussed. A modern approach to the hole measurement is proposed, based on the automatic extraction of geometric data from CAD models, along with the concept of automatic measurement tool selection using a multi-criteria tool selection matrix. Potential directions for the further development of the proposed methods are also discussed, especially considering the use of machine learning for the prediction of measurement uncertainty and the optimization of measurement strategy selection. The presented approach represents a novel and original contribution to the advancement of geometric measurement automation methods. The article emphasizes the growing importance of artificial intelligence based tools in coordinate metrology, highlighting the necessity of proper design and collection of measurement data in the form of databases, which serve as the foundation for the effective operation of intelligent algorithms. Key directions for the development of metrology are also indicated, and current trends that are expected to shape this field in the coming years are described.

Keywords: Cylindrical Hole · CMM · Optical Coordinate Measurement · Artificial Intelligence · Point Cloud

M. Hryb et al. (Eds.): MANUFACTURING 2026, LNME, pp. 233–247, 2026.
https://doi.org/10.1007/978-3-032-23038-6_17

1 Introduction

With the development of Industry 4.0 and 5.0 concepts, rapid progress has been observed in the field of metrology and measurement systems [1–3]. Particular emphasis is placed on optimizing individual stages of production [4], including measurement processes, as well as effective planning. As noted by Solano et al. [5], in process planning, it is essential to verify whether the resources selected for each stage of the plan have sufficient capabilities.

1.1 Cylindrical Hole Measurement – Current Approaches and Challenges

Cylindrical hole measurements remain a significant challenge in metrology. Various coordinate measurement systems are used in this process. The authors of [6] compared selected systems in terms of their capabilities for hole measurements. In optical systems, technologies such as structured light, laser triangulation, and tomography are commonly applied. The structured light method allows for efficient measurement of large-scale objects, providing fast and accurate surface representation. In production environments, robots equipped with triangulation probes are widely used, typically offering higher measurement precision compared to structured light systems. Tomography, on the other hand, enables very precise measurements with the ability to penetrate the interior of objects, although it is associated with higher operational and implementation costs.

Various research centers and Research and Development departments in industry are actively seeking optimal solutions for hole measurement techniques. For example, in [7], the authors propose an "on-machine" measuring device mounted directly on the machine to measure the internal diameter, straightness, and roundness of holes with a nominal diameter of 100 mm and a depth of 700 mm. Compared to a traditional CMM (Coordinate Measuring Machine), this approach allows measurements over a larger range. Similarly, in [8], the same authors propose a hole measurement method based on a precise servo drive system. A spindle with a laser sensor moves along the axis of the hole, while the component is rotated by a servo motor, enabling the reconstruction of the hole's internal shape. Least squares fitting and a gradient algorithm were used for data analysis. In [9], the authors proposed a method for visual measurement of deep holes based on a radial Laplace operator, visual noise suppression, and enhancement of hole edge features. A novel algorithm for removing outlier points is presented, allowing accurate detection of holes with a high depth to diameter ratio.

In the case of hole measurements, particular difficulties arise with narrow and deep holes, referred to as HAR (High Aspect Ratio), i.e., holes with a high depth to diameter ratio. In this context, the term describes a geometry in which the hole depth significantly exceeds its diameter, complicating access for measurement tools and increasing the risk of measurement errors. This issue has been analyzed in studies [10–12], where the authors highlighted that, in contact measurements, there is a risk of the probe tip becoming stuck inside the hole. Using a probe that is too thin and long may lead to deflection, whereas increasing the probe stiffness results in higher measuring forces, which can cause material damage.

The search for increasingly efficient solutions is not limited to mechanical issues or data acquisition alone. The objective of the article is to present the elements necessary for developing a multi-criteria tool selection matrix, including the assessment of the capabilities of selected optical coordinate measurement systems for hole measurement. An additional objective is to understand the relationship between measurement accuracy and hole depth. All acquired knowledge is intended to enable more accurate automated hole measurements. Another goal is to improve the stage of the measurement process in which the operator analyzes the data, by automating it using a script for the automatic detection of cylindrical hole parameters in three-dimensional models stored in STEP format. The automation of this process constitutes an essential component of the multi-criteria tool selection matrix.

Working with CAD models is a common element across all measurement techniques. In [13], the proposed methodology utilizes information from a CAD file and the capabilities of CMMs to optimize the measurement process. CAD models play a key role in the evaluation of results in the form of point clouds, giving these considerations broad application potential. The ability to automatically extract geometric data from CAD files, combined with the developed multi-criteria tool selection matrix, provides a significant means for optimizing and designing hole measurement strategies. These considerations are part of the ongoing work within the LIDER15/0306/2024 project, "An Expert system based on machine learning as a new tool for modeling the uncertainty of coordinate measurement," supported by the National Centre for Research and Development under the LIDER program. The project is focused on developing a tool for rapid and simple estimation of measurement uncertainty based on archival data, prior to the actual measurement. This estimation will subsequently be used in a system for selecting the appropriate measurement tool for a given task. The developed solution will enable the selection of a measurement tool for a hole with a specified uncertainty during the measurement planning stage. The authors found comprehensive publications demonstrating the capabilities of optical systems, presenting a wide range of methods along with descriptions of their advantages and limitations [14–16]. However, no solution was identified that would support the decision-making process for selecting a measurement tool for holes.

2 Materials and Methods

In order to automate the cylindrical hole measurement process, it is necessary to understand certain regularities accompanying the measurements and translate them into defined mechanisms. The research was conducted at the Laboratory of Coordinate Metrology of Cracow University of Technology.

The study focused on:

- understanding the relationship between measurement accuracy and hole depth (Sects. 2.1 and 3),
- assessing the capabilities of selected optical coordinate measurement systems for hole measurement tasks, and constructing a tool selection matrix for specific measurement tasks (Sects. 2.2 and 3),
- developing an auxiliary script for the automatic detection of cylindrical hole parameters in 3D models stored in STEP format (Sects. 2.3 and 3).

The analysis of the relationship between measurement accuracy and hole depth was conducted for a reference ring of 38 mm and a hole of 9.5 mm (see Fig. 1a). Measurements providing input data for the multi-criteria tool selection matrix for specific measurement tasks were performed using a specially designed reference artifact in the form of a prism with holes (see Fig. 1b). The artifact was fabricated using additive manufacturing – stereolithography (SLA), with a Formlabs Form 3B+ 3D printer and Grey resin. It can be expected that standards produced using additive manufacturing may exhibit lower accuracy compared to conventional standards; nevertheless, the authors minimized the influence of manufacturing precision by performing thorough calibrations on more precise machines, Zeiss O-Inspect and Leitz PMM 12106, with the calibration data serving as reference values.

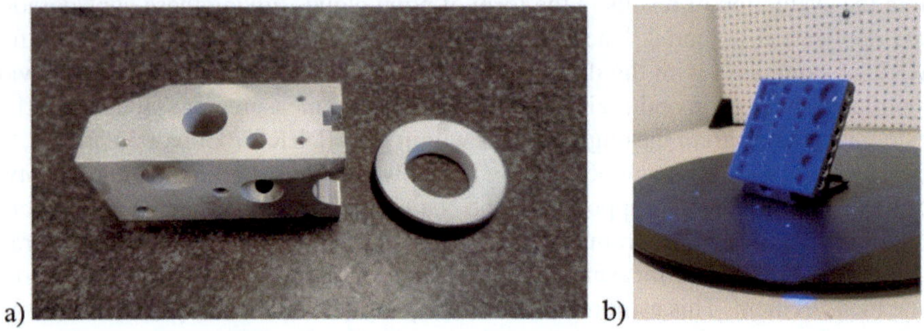

Fig. 1. a) Reference ring of 38 mm and a 9.5 mm hole, b) Prism-shaped artifact with holes fabricated using SLA technology

2.1 Measurement Accuracy of Hole Diameter Along the Hole Axis – Concept Considering an Assumed Segmentation of the Hole into Sequences

The first stage of the study focused on understanding the relationship between measurement accuracy and hole depth. Prior to measurement, the objects were coated with a mattifying spray to eliminate light reflections. The influence of the coating process on measurement data quality is discussed in detail in [17, 18]. Measurements were performed times for each characteristic using a articulated arm coordinate measuring machine equipped with a triangulation probe (see Fig. 3a), and the resulting point clouds were analyzed in GOM Inspect software. The analysis involved constructing cross-sections along the cylinder axis to assess changes in the cylinder diameter with depth, along the hole axis. Areas of constant height (2 mm) were evaluated, excluding the upper spherical zone, since as illustrated in Fig. 5 the points in this region are partially affected by reflections and laser beam scattering at the hole edge. The results are presented in Sect. 3.

2.2 Multi-Criteria Tool Selection Matrix for a Given Measurement Task – Assessing the Capabilities of Optical Systems in Hole Measurements

2.2.1 Matrix Concept

The concept of a multi-criteria tool selection matrix combined with a parametric approach was presented in [19]. This approach is based on collecting fundamental information regarding specific measurement tasks performed on various CMM systems in a database, which is then used as a basis for selecting the appropriate tool for a given measurement task conducted on a particular system. As illustrated in Fig. 2, once a measurement task is unambiguously identified and characterized, the system suggests possible task execution options based on multiple criteria. For one user, measurement uncertainty may be the key criterion, while for another, time or cost may take precedence. Uncertainties in the databases will be estimated in accordance with the GUM (Guide to the expression of uncertainty in measurement) requirements [20]. The matrix also includes mandatory criteria that must be satisfied, such as compliance with the measurement range offered by the machine. Advanced matrices can additionally be equipped with more sophisticated modules for controlling workshop operations, including the availability of a given device at the measurement station.

The difference between the matrices developed at the Laboratory of Coordinate Metrology and the catalog data provided by manufacturers lies in the fact that the proposed solutions are based on archival data from specific measurement tasks. They identify matches between the analyzed task and tasks in the database, and based on this, select the appropriate measurement tool for the given application.

For input data in the form of a description of a given task that differs from the tasks in the database, tool selection will be supported by artificial intelligence. It will assist in determining the degree of similarity between tasks and in finding solutions that are as closely matched as possible.

An example of its application is the scenario mentioned in the article: when a user needs to measure a hole, they specify the expected size, the material of the object, the desired measurement accuracy and tolerances, and optionally choose the type of system. The operator will be able to specify the criteria themselves, indicating whether accuracy, time, or cost is the most important. The matrix will provide the user with a result in the form of tools ranked in order of their suitability for the intended application.

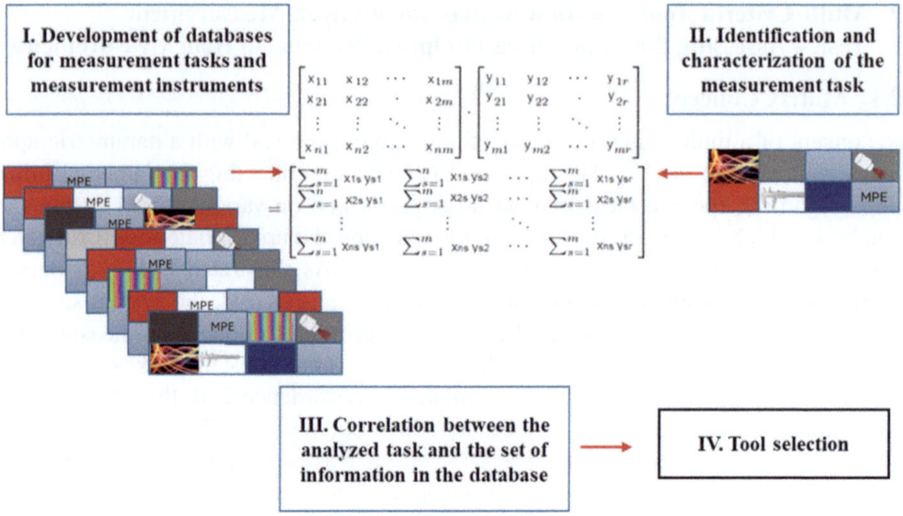

Fig. 2. Concept of the multi-criteria tool selection matrix

2.2.2 Acquisition of Measurement Data for the Matrix

The data used to build the matrix were obtained from four coordinate measurement systems (see Fig. 3): ROMER Absolute Arm RA-7320 SI AACMM with a laser triangulation probe, with an accuracy of approximately ±38 μm, Phoenix V|tome|x M M240 computed tomography system from Waygate Technologies, with an accuracy of SD = (3.8 + L/100 μm), EINSCAN-SE 3D scanner Shining 3D, ATOS Core 300 3D scanner.

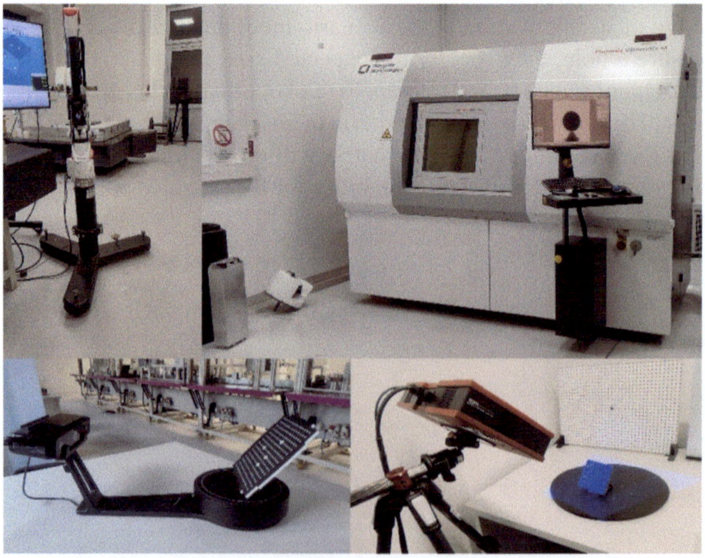

Fig. 3. a) AbsoluteArm RA-7320 SI; b) Phoenix V|tome|x M300 computed tomography system; c) EINSCAN-SE 3D scanner; d) ATOS Core 300 3D scanner

The detailed process of acquiring measurement data is described in [6]. These measurements were supplemented with data obtained using a structured light scanner operating in Triple Scan mode. The component a plate with holes (see Fig. 1b) was measured ten times, resulting in point cloud data from which the diameters and depths of the holes, as measurable by the selected systems, were determined.

2.3 Construction of the CAD Model

The measurement data of the plate obtained using the Triple Scan technology were processed in Geomagic Design X following the sequence shown in Fig. 4:

1. Imported the data into the software in *stl format.
2. Positioned the model in the workspace.
3. Cleaned the model from measurement noise by removing points from reflections.
4. Performed surface region segmentation, including both automatic and manual separation of areas with similar geometric features, such as planes and cylinders.
5. Fitted geometric elements to the identified regions, assigning appropriate analytical shapes (e.g., planes, cylinders, holes) to create a parametric description of the geometry.
6. Created the CAD model after detecting and combining all regions.

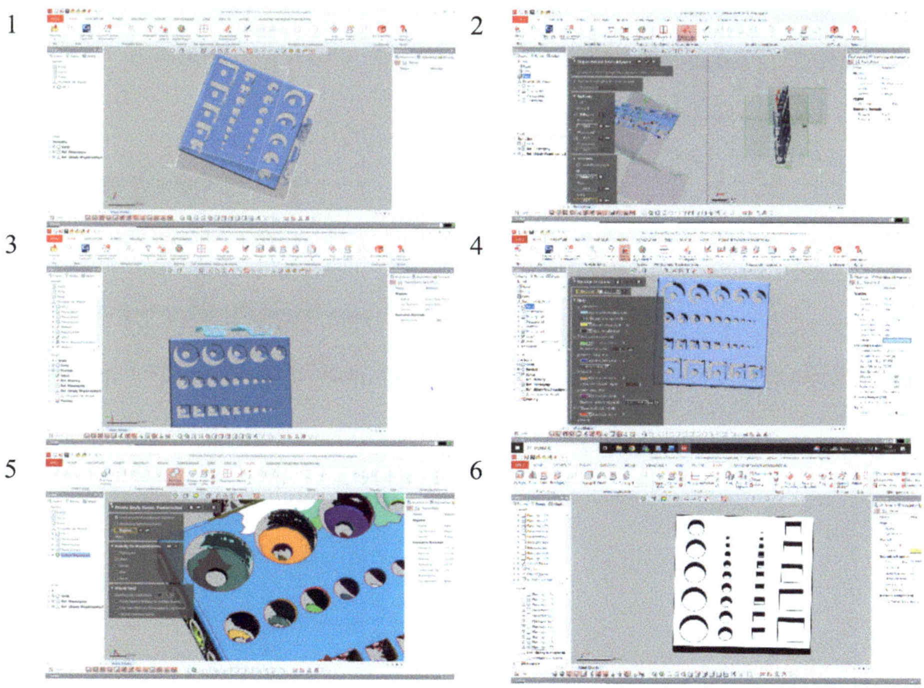

Fig. 4. Construction of the CAD Model in Geomagic Design X

After completing the geometry reconstruction process in Geomagic Design X, a CAD model of the object was obtained. As part of further work directed at automating the hole measurement process, a Python script was developed to automatically detect the parameters of cylindrical holes in three-dimensional models stored in STEP format (Sect. 3).

3 Results

Example measurement results of the cylinder holes are shown in Fig. 5. The cylinders were intentionally designed without the top surface, as, illustrated in Fig. 5, points from this region would distort the results. An increasing trend in diameter values can be observed. As the measurement progresses deeper into the hole, the device records data that result in diameter values increasing by hundredths and thousandths of a millimeter.

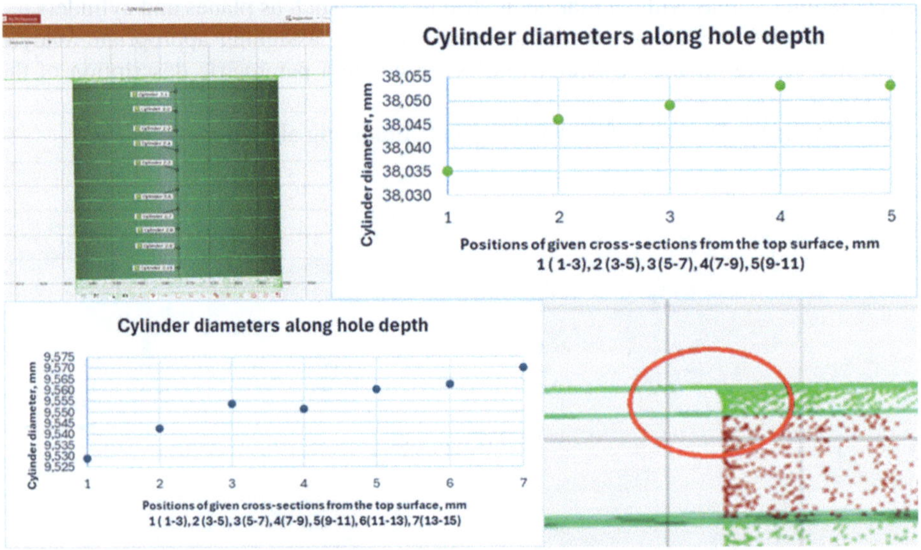

Fig. 5. Measurement results of cylindrical hole diameters at different depths along the hole axis

The tables below (Tables 1, 2, 3, and 4) present the results required for the development of the multi-criteria tool selection matrix.

Table 1. Measurement results of the cylinder diameter for the laser triangulation and computed tomography [6]

No	Nominal diameter, mm	Laser triangulation			Computed tomography (CT)		
		$\bar{x}$, mm	Δx, mm	δ, %	$\bar{x}$, mm	Δx, mm	δ, %
1	20.0614	19.984	0.077	0.386	20.024	0.037	0.185
2	18.0479	17.958	0.090	0.498	18.033	0.015	0.081
3	16.0285	16.129	0.101	0.627	16.011	0.018	0.112
4	14.0100	13.973	0.037	0.264	13.989	0.021	0.153
5	12.0295	12.098	0.069	0.569	12.001	0.028	0.235
6	10.2404	10.234	0.006	0.062	10.193	0.047	0.463
7	9.0198	9.006	0.014	0.153	9.085	0.065	0.723
8	8.0312	8.002	0.029	0.364	7.989	0.043	0.530
9	7.0270	6.972	0.055	0.783	6.991	0.036	0.514
10	6.0146	6.109	0.094	1.570	5.979	0.035	0.585
11	5.0183	4.905	0.113	2.258	4.973	0.046	0.911
12	3.8661	3.901	0.035	0.903	3.940	0.074	1.904
13	2.9436	2.777	0.167	5.660	2.959	0.015	0.516
14	1.9641	–	–	–	1.934	0.030	1.538
15	0.9544	–	–	–	0.891	0.064	6.664
16	0.5886	–	–	–	–	–	–
17	0.3401	–	–	–	–	–	–

Table 2. Measurement results of the cylinder diameter for the structured light

No	Nominal diameter, mm	Structured light			Structured light (Triple scan)		
		$\bar{x}$, mm	Δx, mm	δ, %	$\bar{x}$, mm	Δx, mm	δ, %
1	20.0614	20.086	0.025	0.123	20.357	0.296	1.473
2	18.0479	18.153	0.105	0.582	18.228	0.180	0.998
3	16.0285	15.923	0.106	0.658	16.218	0.189	1.182
4	14.0100	13.982	0.028	0.200	14.227	0.217	1.549
5	12.0295	12.094	0.064	0.536	11.957	0.072	0.603
6	10.2404	10.163	0.077	0.756	10.047	0.193	1.889
7	9.0198	9.004	0.016	0.175	9.006	0.014	0.153
8	8.0312	8.098	0.067	0.832	8.026	0.005	0.065
9	7.0270	7.095	0.068	0.968	6.996	0.031	0.441
10	6.0146	6.084	0.069	1.154	5.950	0.065	1.074

(*continued*)

Table 2. (*continued*)

No	Nominal diameter, mm	Structured light			Structured light (Triple scan)		
		$\bar{x}$, mm	Δx, mm	δ, %	$\bar{x}$, mm	Δx, mm	δ, %
11	5.0183	5.167	0.149	2.963	4.865	0.153	3.055
12	3.8661	3.802	0.064	1.658	3.865	0.001	0.028
13	2.9436	2.510	0.434	14.730	2.835	0.109	3.689
14	1.9641	–	–	–	1.832	0.132	6.721
15	0.9544	–	–	–	–	–	–
16	0.5886	–	–	–	–	–	–
17	0.3401	–	–	–	–	–	–

Table 3. Measurement results of the cylinder depth for the laser triangulation and computed tomography [6]

No	Nominal depth, mm	Laser triangulation			Computed tomography (CT)		
		$\bar{x}$, mm	Δx, mm	δ, %	$\bar{x}$, mm	Δx, mm	δ, %
1	10.0300	10.005	0.025	0.250	10.012	0.018	0.180
2	10.0320	10.066	0.034	0.336	10.058	0.026	0.260
3	10.0150	10.005	0.010	0.098	10.071	0.056	0.556
4	10.0430	10.014	0.029	0.290	10.041	0.002	0.025
5	10.0290	9.988	0.041	0.414	9.990	0.039	0.387
6	10.0320	10.004	0.028	0.277	9.987	0.045	0.449
7	10.0070	9.989	0.018	0.176	10.015	0.008	0.075
8	10.0390	9.993	0.046	0.457	9.957	0.082	0.821
9	10.0330	10.003	0.030	0.299	10.018	0.015	0.152
10	10.0560	10.022	0.034	0.341	10.006	0.050	0.496
11	10.0700	–	–	–	10.051	0.019	0.190
12	10.0430	–	–	–	9.935	0.108	1.073
13	9.9560	–	–	–	9.907	0.049	0.496
14	8.1180	–	–	–	9.961	1.843	22.705
15	–	–	–	–	3.720		
16	–	–	–	–	–	–	–
17	–	–	–	–	–	–	–

Table 4. Measurement results of the cylinder depth for the structured light

No	Nominal depth, mm	Structured light			Structured light (Triple scan)		
		$\bar{x}$, mm	Δx, mm	δ, %	$\bar{x}$, mm	Δx, mm	δ, %
1	10.0300	9.544	0.486	4.845	9.700	0.330	3.290
2	10.0320	9.479	0.553	5.512	9.988	0.044	0.439
3	10.0150	9.242	0.773	7.718	10.015	0.000	0.000
4	10.0430	8.931	1.112	11.072	9.868	0.175	1.743
5	10.0290	8.509	1.520	15.156	9.618	0.411	4.098
6	10.0320	9.778	0.254	2.532	9.725	0.307	3.060
7	10.0070	9.757	0.250	2.498	9.921	0.086	0.859
8	10.0390	9.387	0.652	6.495	10.148	0.109	1.086
9	10.0330	–	–	–	10.034	0.001	0.010
10	10.0560	–	–	–	–	–	–
11	10.0700	–	–	–	–	–	–
12	10.0430	–	–	–	–	–	–
13	9.9560	–	–	–	–	–	–
14	8.1180	–	–	–	–	–	–
15	–	–	–	–	–	–	–
16	–	–	–	–	–	–	–
17	–	–	–	–	–	–	–

Further analysis of measurement accuracy versus hole depth is planned, based on a more diverse sample of hole diameters.

As an additional component of the work, a Python script was developed to automate the process of extracting data from CAD models. Its purpose is to detect the parameters of cylindrical holes in three-dimensional models stored in STEP format. The script was implemented in the CADQuery environment using the Open Cascade (OCP) library interface, which provides direct access to the topological and geometric information of model elements. Upon execution, the script loads the STEP model and generates a list of detected holes, including their diameters, depths, and tolerances. The analysis results are presented in a textual format, allowing comparison with measurement data. Example records of data extraction from a CAD model:

Hole 1: Diameter = 16.000 ± 0.200 mm, Depth = 10.000 ± 0.200 mm
Hole 2: Diameter = 12.000 ± 0.200 mm, Depth = 10.000 ± 0.200 mm.

4 Discussion

As part of the analyses directed at understanding the relationship between measurement accuracy and hole depth, it was found that for a hole with a diameter of 9.5 mm, measurement data could be obtained to a depth of approximately 15 mm. A similar measurement for a hole with a diameter of 20 mm allowed point registration up to a depth of 33 mm. Based on these results, a preliminary relationship can be cautiously formulated, indicating that when using a laser triangulation system, the feasible depth to diameter measurement ratio of a hole is approximately: **1.7:1**. It should be emphasized, however, that further research is planned to confirm and refine this relationship.

In the next stage of the study, aimed at evaluating the capabilities of selected optical coordinate measurement systems for hole measurement tasks, it was determined thatlaser triangulation is no longer capable of providing reliable measurements when measuring diameters of 2 mm. In general, diameters below 6 mm fall outside the accuracy limits of the measuring device as specified by the manufacturer. This limitation is related to the system design: the camera is positioned at an angle relative to the light source, and even if the laser line is visible on the object to the human observer, the camera may not detect it because its field of view is occluded by the edge of the measured object. Computed tomography (CT), on the other hand, loses measurement reliability only at diameters of approximately 1 mm, which may be associated with the voxel size (Table 1).

The stereo structured light system (with two cameras mounted at an angle relative to each other), similarly to laser-based systems, is not capable of measuring internal cylinders with diameters below 2–3 mm. Likewise, as in the previously discussed system, all diameters below 6 mm fall outside the manufacturer's specified accuracy range. These observations indicate that similarities in the design of these systems result in comparable measurement performance. In devices where an additional camera is integrated into the projector positioned between the two cameras, one additional measurable result was observed, allowing measurement down to a cylinder diameter of 2 mm. However, measurements below 5 mm still fall outside the manufacturer's specifications. The applied technology increased the capability to reproduce such diameters by approximately 1 mm (Table 2).

In the test designed to determine the minimum internal cylinder diameter for which a depth of approximately 10 mm could be measured, laser triangulation produced measurements within the manufacturer's specified accuracy down to a cylinder diameter of 6 mm, whereas computed tomography remained reliable down to a diameter of 2 mm. The stereo structured light system was able to measure the same depth only down to a cylinder diameter of 8 mm. In devices with the additional camera integrated into the projector, one additional measurable result was observed, enabling depth measurement down to an internal cylinder diameter of 7 mm (Table 3).

These results collectively demonstrate that the limitations in minimum measurable diameters are strongly influenced by the system design, camera placement, and measurement technology. Systems with structural or design enhancements, such as additional cameras, provide slight improvements, yet measurements below the manufacturer-specified limits remain unreliable (Table 4).

Thus, when constructing the multi-criteria tool selection matrix, an operator tasked with measuring a small hole and having access to the mentioned systems would receive a recommendation to use computed tomography, provided that the cost criterion is met. The multi-criteria nature of the matrix considers various factors, primarily the feasibility of performing a given task, but also cost and time. Therefore, once the user's preferences are identified, they receive an optimal recommendation tailored to maximize the suitability of the selected measurement tool for their needs.

The final stage, involving the development of an auxiliary script for automatically detecting cylindrical hole parameters in three-dimensional models saved in STEP format, was also successfully completed. The system automatically extracts data from CAD files, significantly enhancing the efficiency of the multi-criteria tool selection matrix and reducing response time. This stage will considerably accelerate the CAD data extraction process.

The task of developing a multi-criteria tool selection matrix is highly complex, which necessitates focusing not only on the construction and usability of the matrix itself but also on reasoning based on the known principles of hole measurement and the capabilities of the measurement systems. This explains the diversity of the presented stages of work and analyses.

The developed tool offers a range of advantages. First and foremost, it aligns with the principles of automating decision-making processes and implementing intelligent solutions. Process automation reduces the likelihood of errors by less experienced operators and accelerates the measurement process. In the context of tool selection, the matrix minimizes selection errors and, in the case of purchases, reduces the risk of acquiring a device whose capabilities are not properly matched to specific tasks. It also enables the simulation of measurement paths for tasks that are still in the design phase. The project contributes to reducing energy consumption by providing better knowledge of the measurement process before it is executed, allowing the overall process to be optimized. A limitation of the solution is that such database-based tools rely on extensive information, and their usefulness is closely linked to the need for conducting a series of tests. It should also be noted that these solutions are based on statistical averages, so machines – even from the same manufacturer – may differ when located in different laboratories. Nevertheless, the analysis concluded that the benefits significantly outweigh the drawbacks, and it was decided to continue work in this direction.

5 Conclusions

The presented research and analyses enabled the determination of the relationship between the accuracy of optical coordinate measurement and hole depth for the investigated case. The capabilities of selected optical coordinate measurement systems for hole measurement tasks were assessed, which allowed the acquisition of initial data for the construction of a multi-criteria tool selection matrix for specific measurement tasks.

It was determined that computed tomography (CT) enabled the mapping of the largest number of cylinder diameters, including small-diameter cylinders, e.g., 1.934 mm. It should be noted that each system exhibited a different critical point regarding decreasing diameters at which further accurate mapping was no longer possible. When analyzing cylinder depth measurements, the fewest results were obtained using structured light systems. Fixing the object on a rotary table restricted the free maneuvering of the light beam, limiting access to hard-to-reach areas. In this scenario, laser triangulation proved more efficient. In the last step an auxiliary script was developed to automatically detect cylindrical hole parameters in three-dimensional models saved in STEP format. All conducted work confirmed the feasibility of automating the hole measurement process, both at the stage of planning the measurement task and during data analysis. The developed matrix will also integrate the Expert system based on machine learning, serving as a new tool for modeling the uncertainty of coordinate measurement. By combining all these elements, the operator will have a comprehensive tool enabling them to plan hole measurements with a defined uncertainty at the task design stage – that was previously not possible. Since the developed system – the multi-criteria tool selection matrix, operates based on databases, it is planned to be enhanced with machine learning components, providing the operator with an advanced tool aligned with the principles of Industry 4.0 and 5.0.

Acknowledgements. This study was supported by the National Center for Research and Development, Poland, under the LIDER program (grant no. LIDER15/0306/2024). The authors also acknowledge the Laboratory of Coordinate Metrology for providing access to the measurement facilities.

References

1. Stojadinovic, S.M., Majstorovic, V.D., Gąska, A., Sładek, J., Durakbasa, N.M.: Development of a coordinate measuring machine – based inspection planning system for industry 4.0. Appl. Sci. **11**, 841 (2021)
2. Catalucci, S., et al.: Smart optical coordinate and surface metrology. Meas. Sci. Technol. **34**(1), 012001 (2022)
3. Taatali, A., Sadaoui, S.E., Louar, M.A., Mahiddini, B., Catalucci, S., Savio, E.: A novel study on camera calibration and deep learning based processing techniques for efficient machine vision inspection. Meas. Sci. Technol. **36**(8)
4. Kowalczyk, M., Tomczyk, K.: Procedure for determining the uncertainties in the modeling of surface roughness in the turning of NiTi alloys using the Monte Carlo method. Materials. **13**(19), 4338 (2020)
5. Solano, L., Rosado, P., Romero, F.: Ontological-based validation of selected technological resources in integrated machining and inspection process planning. Procedia Eng. **132**, 1096–1103 (2015)
6. Ostrowska, K., Owczarek, D., Jędrzejak, I., Felho, C.: Comparative analysis of the capabilities of reverse engineering systems in reconstruction the geometry of machined components. In: Development in Machining Technology: Scientific-Research Reports, pp. 33–42. Monograph Cracow University of Technology (2025)
7. Liang, J., Song, X., Wang, K., Han, X.: An on-machine measuring apparatus for dimension and form errors of deep-hole parts. Sensors. **24**(23), 7847 (2024)
8. Liang, J., Wang, K., Song, X., Han, X.: Interior profile accuracy assessment method of deep-hole parts based on servo drive system. Sensors. **24**(20), 6554 (2024)
9. Meng, F., et al.: A visual measurement method for deep holes in composite material aerospace components. Sensors. **24**(12), 3786 (2024)
10. Hovell, T., et al.: Measurement techniques for three-dimensional metrology of high aspect ratio internal features—a review. Metro. **3**, 139–168 (2023)
11. Murakami, H., Katsuki, A., Onikura, H., Sajima, T., Kawagoishi, N., Kondo, E.: Development of a system for measuring micro hole accuracy using an optical fiber probe. J. Adv. Mech. **4**, 995–1004 (2010)
12. Meli, F., Küng, A.: AFM investigation on surface damage caused by mechanical probing with small ruby spheres. Meas. Sci. Technol. **18**(2), 496 (2007)

13. Chouridis, I., Mansour, G., Papageorgiou, V., Mansour, M.T., Tsagaris, A.: Measurement optimization from CAD using artificial intelligence. Metro. **5**(4), 64 (2025)
14. Zhao, F., Tang, H., Zou, X., Li, X.: A review of optical metrology techniques for advanced manufacturing applications. Micromachines. **16**(11), 1224 (2025)
15. Shan, S., Zhao, F., Li, Z., Luo, L., Li, X.: A comprehensive review of optical metrology and perception technologies. Sensors. **25**(22), 6811 (2025)
16. Huang, G., et al.: A review of optical interferometry for high-precision length measurement. Micromachines. **16**(1), 6 (2025)
17. Owczarek, D., et al.: The impact of an object's surface material and preparatory actions on the accuracy of optical coordinate measurement. Materials. **18**, 3693 (2025)
18. Franke, J., Koutecký, T., Koutný, D.: Comparison of sublimation 3D scanning sprays in terms of their effect on the resulting 3D scan, thickness, and sublimation time. Materials. **16**(18), 6165 (2023)
19. Owczarek, D.: Uncertainty Model in Optical Coordinate Measurement. Monograph Cracow University of Technology, Cracow (2017)
20. Joint Committee for Guides in Metrology: Evaluation of measurement data – guide to the expression of uncertainty in measurement. JCGM. **100**, 2008 (2008)

Author Index

M. Hryb et al. (Eds.): MANUFACTURING 2026, LNME, pp. 249–250, 2026.
https://doi.org/10.1007/978-3-032-23038-6